Biodesign in the Age of Artificial Intelligence: Deep Green investigates the potential of nature-based technology for shaping the evolution of contemporary architecture and design. It takes on the now pervasive topic of design intelligence, extending its definition to encompass both biological and digital realms.

As in their first title, *Systemic Architecture: Operating Manual for the Self-Organizing City*, the authors engage the topic through the specific lens of their innovative design practice, ecoLogicStudio, and their research at the University of Innsbruck and at the Bartlett, UCL. Part One of the book, entitled PhotoSynthetica™, illustrates design solutions that engage the urban microbiome and seek to achieve an immediate impact, while Part Two, entitled Deep Green, includes synthetic landscapes and operates within a much larger spatio-temporal frame, going beyond human perception and life span to envision design as a geographical and geological force.

In the age of catastrophic climate change, such perceptual expansion helps to clarify that change cannot simply be stopped or rolled back. We must instead establish more positive dynamics of change within the living world. To this end, this book proposes to engage with design and architecture as an extended cognitive interface, a sentient being that is co-evolutionary and symbiotic with the living planet, contributing to its beauty and to our continued enjoyment of it.

Claudia Pasquero and Marco Poletto are authors, educators and design innovators. They live and work between London and Innsbruck.

Biodesign in the Age of
Artificial Intelligence

DEEP GREEN

Claudia Pasquero & Marco Poletto

Cover image: H.O.R.T.U.S. XL at Centre Pompidou captured by NAARO.

First published 2023
by Routledge
605 Third Avenue, New York, NY 10158

and by Routledge
4 Park Square, Milton Park, Abingdon, Oxon, OX14 4RN

Routledge is an imprint of the Taylor & Francis Group, an informa business

© 2023 Claudia Pasquero and Marco Poletto

The right of Claudia Pasquero and Marco Poletto to be identified as authors of this work has been asserted in accordance with sections 77 and 78 of the Copyright, Designs and Patents Act 1988.

All rights reserved. No part of this book may be reprinted or reproduced or utilised in any form or by any electronic, mechanical, or other means, now known or hereafter invented, including photocopying and recording, or in any information storage or retrieval system, without permission in writing from the publishers.

Trademark notice: Product or corporate names may be trademarks or registered trademarks, and are used only for identification and explanation without intent to infringe.

Library of Congress Cataloging-in-Publication Data
A catalog record for this title has been requested

ISBN: 9780367768041 (hbk)
ISBN: 9780367768010 (pbk)
ISBN: 9781003168614 (ebk)

DOI: 10.4324/9781003168614

Typeset in Helvetica Neue & ARK ES
by WWW Stuudio & Robin Siimann

This book has been prepared from camera-ready copy provided by the authors.

TABLE OF CONTENTS

Preface by Claudia Pasquero and Marco Poletto 2

Introduction by Sir Peter Cook 5

PART 1 PHOTOSYNTHETICA

PhotoSynthetica 10

Projects 1.01
PhotoSynthetica test beds, pilot projects and early adoptions 38

Biotechnological architecture
On culturalising the urban microbiome 110

Projects 1.02
Bio-digital sculptures: designing the living 126

Environmentalism beyond ideology
Reprogramming the blue-green city 178

Projects 1.03
Cyber-gardening the city 192

PART 2 DEEP GREEN

The Polycephalum
A journey through architecture, biology and cognition 230

Projects 2.01
Synthetic landscapes 252

Projects 2.02
Deep planning 304

Tomatoes and the immortality of the soul by Prof. Mario Carpo 337

Biographies 339

Credits 341

Index 347

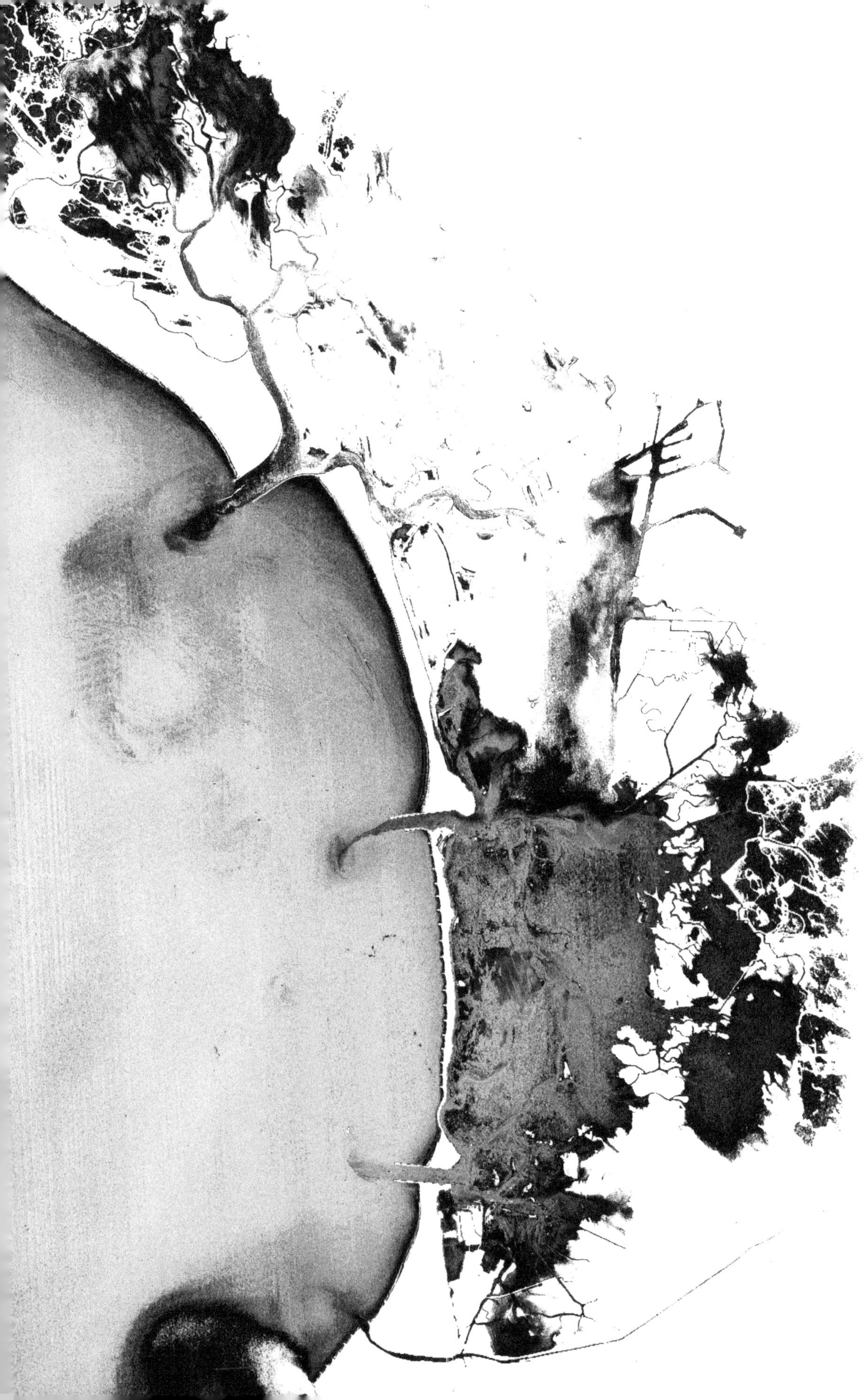

Biodesign in the Age of Artificial Intelligence: Deep Green is our second book about the potential of biotechnology for shaping the evolution of contemporary architecture and design. The focus of this book is the pervasive topic of design intelligence, extending its definition to encompass both biological and digital realms.

As with our previous title, *Systemic Architecture: Operating Manual for the Self-Organizing City*, we engage with the topic through the specific lens of our innovative design practice, ecoLogicStudio. In the 10 years that have elapsed since the publication of *Systemic Architecture*, this practice has evolved into a design innovation consortium, with close ties to internationally renowned academic institutions such as UCL London, the University of Innsbruck and the IAAC in Barcelona. This book illustrates this evolution by presenting a selection of milestone works completed in the period from 2012 to 2022.

Perhaps the most significant of those works is the completion of our practice-based PhDs, from which this book derives most of its theoretical content. That experience led us to a more mature and structured relationship between our academic research and the application of design innovation in our practice, and this has greatly informed the most recent projects of ecoLogicStudio. Ultimately, the purpose of this book is to present a vision of what a venturous architectural practice could look like in the years to come.

In this age of global pandemic, it is perhaps inevitable that the works presented here also reveal a more intimate dimension of our practice, and the growing presence of our two children. Projects like BioBombola, AirBubble and BIT.BIO.BOT reflect upon the role of the family in a creative practice and test how design innovation can enable radical lifestyle choices, individual as well as collective ones. We wish to thank Giacomo and Lulu for their unconditional love and support throughout this uniquely intense period.

The theoretical and technical design innovations presented in this book are inherently practice-based and so would not have been possible without an enabling network of brilliant clients, inspiring colleagues, daring researchers, committed students, honest friends, supportive family, resilient fungi, ubiquitous moulds and proliferating bacteria. These human and non-human interactions constitute a kind of hive mind, a network that is both material and meta-physical, for which resource we are immensely grateful. We owe our thanks to many more people than can be named here, and so we have attempted instead to acknowledge our debt in the credits that accompany each project listed at the end of the book.

At the time of writing there is no doubt in our mind that this network has grown its own very unique collective intelligence, and the projects presented in this book are a testament to our belief that true creativity and innovation are an emergent quality, manifested in the material articulation and aesthetic value of each work. Beauty and pleasure, as enjoyed during the production, completion and sharing of a project are an embedded meta-language that communicates to the engaging minds the project's true ecological intelligence. As such, this book seeks to reaffirm the profound contribution that design practices can make to society's cultural evolution and seeks to surpass the outdated model that understands design practice as separated from and often subservient to theory and critical thinking.

It follows that the notion of intelligence must be central to this book. However, we do not wish to dwell on new definitions of what may constitute design intelligence, artificial or biological. Rather, this book illustrates, especially in its second part, where that intelligence may be found outside the individual human mind, what it may consist of and how it might appear. Our aim is to go beyond the narrow vision of intelligence that is endorsed in the media and in movies, a vision that is often utilitarian, exploitative and profit-seeking in nature.

Instead, this book proposes a concept of artificial intelligence that is more like a slime mould, a spider's web, a microalgae colony or a mycelium network. The architectures and landscapes that are proposed in the following pages embody these forms of intelligence in their morphology, material behaviour or aesthetic appearance. As a consequence, they all propose solutions that do not seek to exhaust the energy and resources they extract from the planet, rather they seek to grow and evolve as a constant regenerative process and what appears to be a new kind of artificial circularity—through the re-metabolisation of waste or the filtration of pollution, for example. In other words, this book proposes a model for an embodied design intelligence. Its subjects are living photosynthetic architectures, bio-digital sculptures and cyber-Gardens in the public realm, as illustrated in the three chapters of Part One, as well as Synthetic Landscapes and the Deep City, which are the principal subjects of Part Two. Together they envision a new technology of nature that seeks to shake us from our anthropocentric indifference to the beautiful intelligence of our living planet.

Following a delightful foreword by Sir Peter Cook, Part 1 of the book, entitled PhotoSynthetica™, proposes design solutions that engage the urban microbiome and seek to achieve an immediate impact. Part 2 of the book, entitled Deep Green, which is masterfully wrapped up by the afterword of Professor Mario Carpo, operates within a much larger spatio-temporal frame, going beyond human perception and life span to envision design as a geographical and geological force.

By now, in the age of catastrophic climate change, it has become clear that change cannot simply be stopped or rolled back. Therefore, we must instead establish a more positive dynamics of change within the living world. To this end, this book proposes to engage with design as an extended cognitive interface, a sentient being that is co-evolutionary and symbiotic with the living planet, contributing to its beauty and to our continued enjoyment of it.

Claudia Pasquero and Marco Poletto.
London, 29 August 2022.

Sir Peter Cook

Introduction

It may be my good fortune to be able to observe a particular circuit of younger friends who have avoided the doom-obsessed condition of many of their contemporaries. Instead— inventing their way forward to face the vicissitudes of Climate and Resource. Forming a little cobweb that criss-crosses from the Architectural Association in London to its rival, the Bartlett (up the street), then across to the University of Innsbruck and then tied together by the Royal Melbourne Institute of Technology University where most of the participants gained their doctorates. Out of such an academic-sounding network might predictably come some bitsy and unfathomable contraptions in a college yard and some dry, clever-sounding reports.

Not so, and particularly not so in the case of ecoLogicStudio. Their stuff is pretty comprehensible, it is definitely out there— as becomes obvious when we take a look at their many installations. It also goes out of its way to be very friendly, not least through Claudia Pasquero and Marco Poletto encouraging their own children to demonstrate the wit of the structures that they build; encouraging members of the public to handle, taste, even manipulate some of the gadgetry and products.

For, in a world of nervousness and suspicion, they are doing 'stuff'. Of course, in the texts accompanying that stuff there are logical, step-by-step procedures, even some dense pieces of reporting that might have crawled out of a scientific verification document and for a (dazed) reader such as myself, some intriguing new terminologies that I hope were invented by themselves. Of these I particularly like the term 'meta-Folly'. It conjures up all sorts of speculative (and creatively cynical) cross-thoughts about the tiresome status of reality in relation to invention, the idea of a diversionary tactic, the value of 'folly' and so on. Another is the concept of 'wetware' [Tallinn Wet City] … leading, again to a lateral evaluation of our armoury of artefacts in the present circumstance. Facing the flood, wetware might very well become the language of survival.

The book proceeds very logically from the gradual establishment of the territories that inspire them, celebrating such as "the human sense of beauty and pleasure as a co-evolutionary system of the mind and its surrounding environment". On another flank they remind us that 'PhotoSynthetica'[™] uses the power of algae to absorb carbon dioxide from the air and thus, if developed, it can make a serious contribution to achieving carbon neutrality in cities. I cite these two territories as examples of the many posits that accompany the work that establish a serious link between vision and problem-solving, all the time underpinning the forward march of their installations. Ever extending, ever more confident in their techniques and, again, very clear in their choice of 'demonstration-instrument'.

Perhaps some of their more pretentious contemporaries will be jealous of the apparent ease with which they seem to be able to use almost any circumstance: a triennale here, a biennale there, an office wall, a playground and (surely one day soon) a substantial piece of urban territory that simply and deliberately enhances a bio-related demonstration. At home, on their apartment wall stands the 'BioBombola': a simple algae-grower—as might be a hi-fi or caged parrot in another's home. As with the hi-fi or the parrot, it can engage by offering both the predictable and the unexpected and thus an ongoing stimulus. Elsewhere are the 'BIT.BIO.BOT', the 'Bio.Tech Hut' or 'photo-bioreactor cladding' [BioFactory]. Currently, they inhabit a large space in London's Building Centre, where, over the space of a few weeks, we can watch and sample the making of a fabrication material derived from algae happening in front of our very eyes.

In the latter part of the book, they move onwards and ever upwards in scale: evolving or tracking urban systems and potential urban morphologies that are analogous to—or even grow with—algae canopies and 'proto-gardens'. They reach forward towards a context that has often proved to be inescapable for the highly sophisticated architect-inventor. Like the rest (including myself) the notion, the understanding and the creative suggestion of 'the city' is irresistible. Naturally it throws them back away from the circumstantial, the playful, the gadgetised and back into an academic discourse. Necessary for the brain, intriguing and clever.

Yet this writer—not too secretly—hopes that they don't get too detached by this from the immensely formidable array of working projects that can continuously edge forward in scale, intensity and application towards an actual piece of bio-city.

PART 1
PHOTO-SYNTHETICA

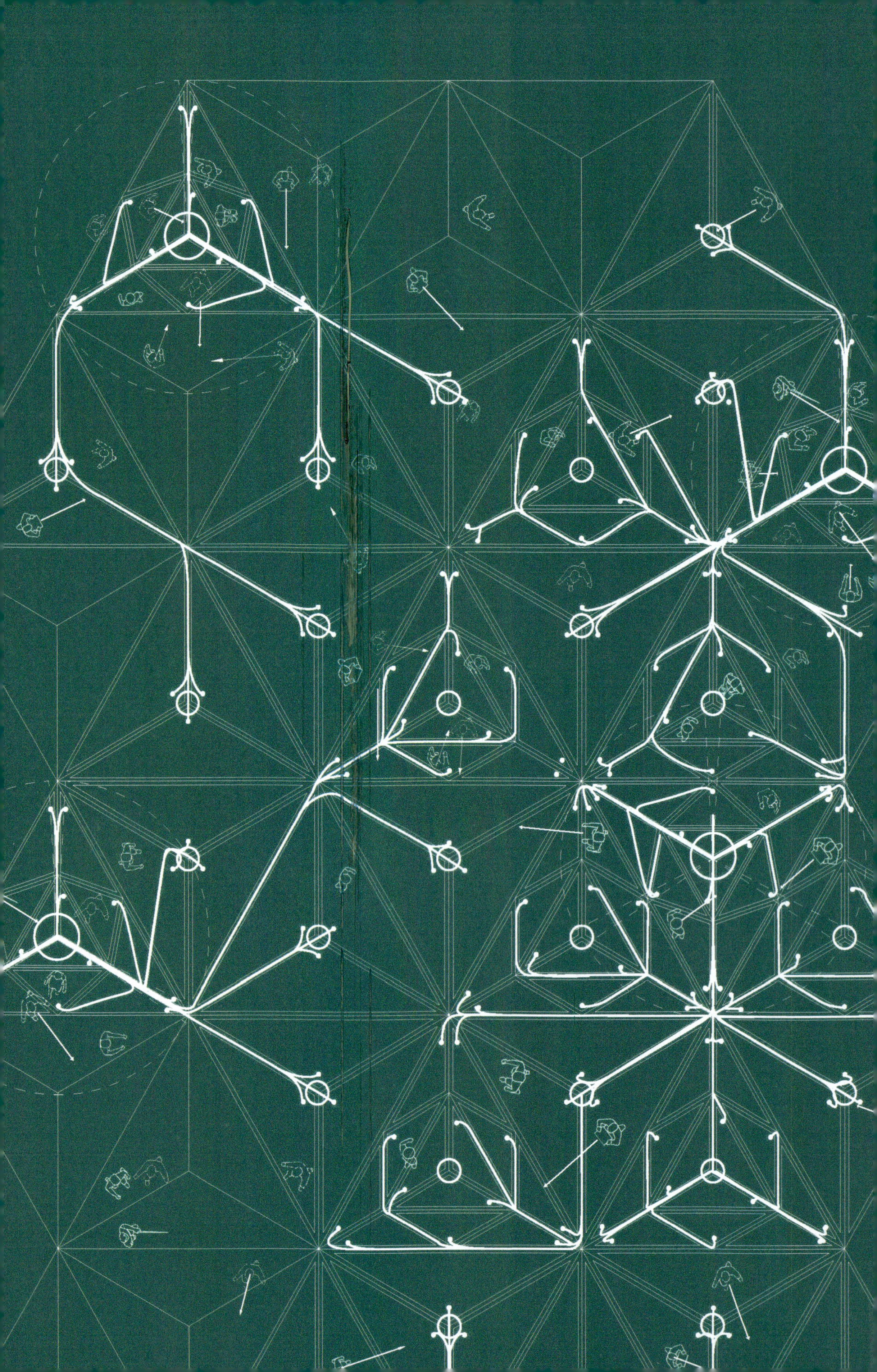

PHOTO-SYNTHETICA

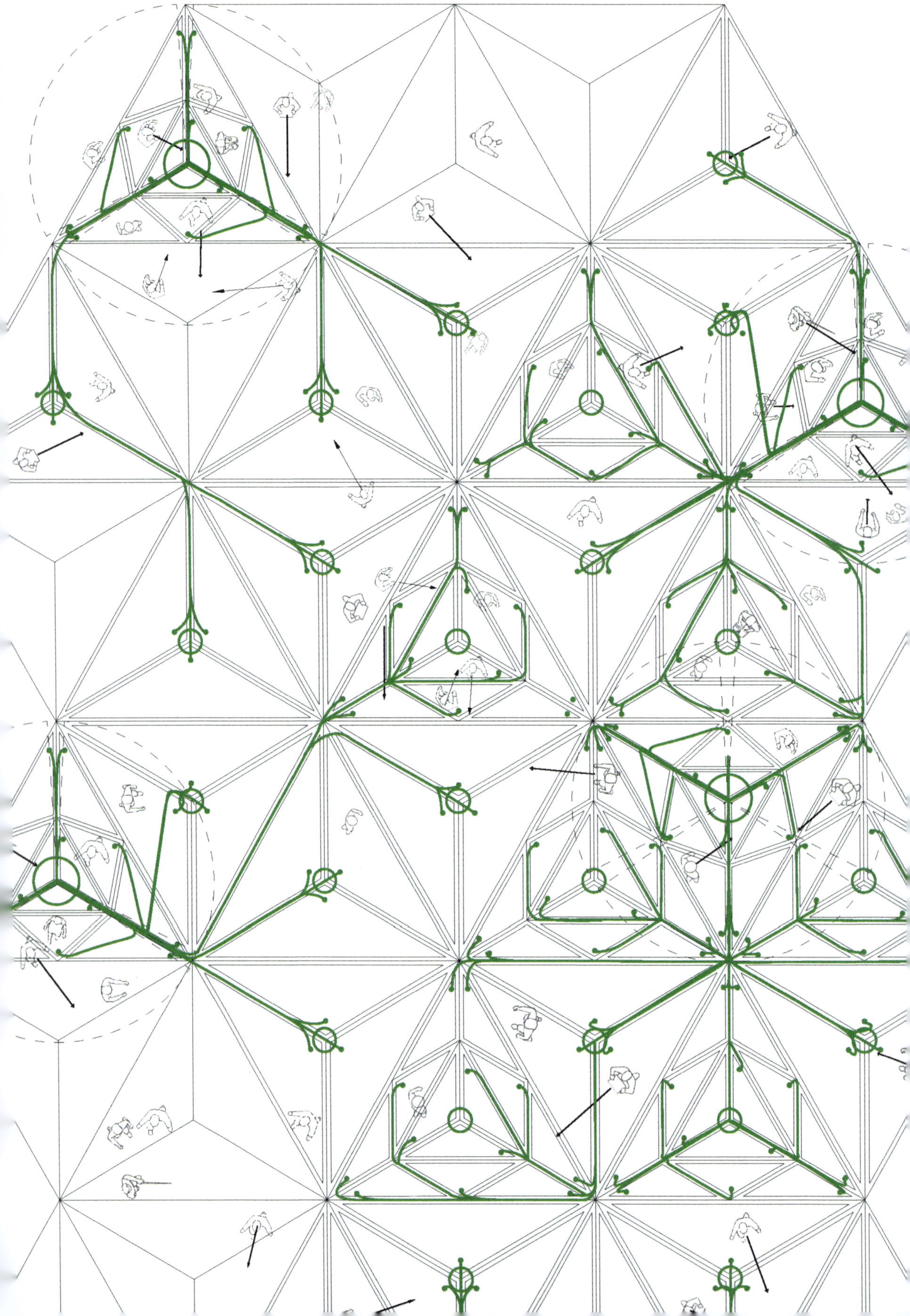

PhotoSynthetica™ uses the power of algae to absorb carbon dioxide from the air. It demonstrates how biotechnology can become integrated in our cities to help achieve carbon neutrality.

PhotoSynthetica™ is a unique biotechnological system for healthy and carbon neutral cities. Powered by solar energy, it harvests the exceptional photosynthetic potential of living microalgae colonies. This is combined with urban emissions monitoring and big data visualisation to efficiently remove both CO2 and pollutants from the urban atmosphere. Filtered particles and gasses are biochemically transformed into biomass, a valuable raw material utilised in several emerging industries.

PhotoSynthetica™ is the most advanced nature-based solution to achieve carbon neutrality, purify the urban atmosphere and create new circular economies. The PhotoSynthetica™ system integrates these objectives in the spaces we inhabit thus enabling a participatory model of urban sustainability.

The PhotoSynthetica™ technology can be adopted in a wide range of urban and architectural scenarios such as: green building facades, vegetated roofs, vertical farms, sky gardens, air purifying public spaces, urban parks, biotechnological playgrounds, vegetated lobbies and courtyards, living walls and screens, air purifying curtains and domestic food kits.

ecoLogicStudio conceived and developed the PhotoSynthetica™ system described in this chapter following a clear set of ethical missions and performative criteria:

Integrating biotechnology in the built environment; decarbonising cities with nature-based technologies; pioneering the bio-smart sector; recognising beauty as a measure of ecological intelligence; converting pollutants into valuable raw material; designing technologies for the bio-conscious city; crafting bio-material solutions; envisioning ecosystemic urban growth; enabling the cyber-gardening revolution; cultivating the public realm collectively.

The challenges

[1] Dasgupta, S., Lall, S., & Wheeler, D. (2021). Urban CO2 Emissions: A Global Analysis with New Satellite Data. Policy Research Working Paper No. 9845. Washington, DC: World Bank. © World Bank. Licence: CC BY 3.0 IGO. https://openknowledge.worldbank.org/handle/10986/36557

Cities emit 70% of global CO_2[1] while buildings alone consume about 36% of the world's primary energy and are responsible for 40% of global urban air pollution. More than 50% of the global urban population breathes air with illegal levels of toxicity and eats unhealthy food. Nevertheless, the total urban population is expected to grow to 6.5 billion by 2050 and the building floor area by a further 60%.

The IPCC has warned that in the same period society must become carbon neutral to avert catastrophic climatic impact. This target can only be achieved if in the next 30 years we transform our buildings into active carbon sinks, our cities into pollution filters and our urban infrastructures into living ecologies able to recirculate waste into valuable raw material. In other words, we need to envision, engineer and produce the architecture of carbon neutrality.

At COP26, in November 2021 in Glasgow, the UK and other governments, together with most industry leaders, have pledged to achieve carbon neutrality by 2050. London, among several other global cities, has signed the Net Zero Carbon Buildings Commitment and has included carbon, clean air and urban greening policies in its new Plan. Achieving these targets will require significant carbon sequestration as well as dramatic reduction in CO_2 emissions.

Currently, the only nature-based means of carbon sequestration are planting more trees, green roofs, and vertical gardens. These solutions are often inefficient and require high maintenance, especially in dense urban environments. Large trees have limited capacity for architectural integration, and their photosynthetic efficiency is greatly impaired by building density.

PhotoSynthetica™ provides carbon sequestration 10 times more efficiently than large trees as algae are microorganisms with cells that are entirely photosynthetic. It also provides an integrated protocol for reducing CO_2 emissions on site, thus achieving a substantial net reduction balance.

As the world population increases, more people live in cities and urban air quality decreases. This makes filtering air pollution in the urban realm a critical challenge. In London for instance, two million people are exposed to illegal levels of toxic pollutants in the air and a large percentage of these people are children. Research has widely demonstrated how children's lungs may be scarred for life if exposed to fine particulate matter (PM2.5) in the developmental age.[2]

[2] US Environmental Protection Agency. (2022). Supplement to the 2019 Integrated Science Assessment for Particulate Matter. Final Report. Washington, DC: US EPA. EPA/635/R-22/028.

Furthermore, prolonged exposure to air pollution weakens the human immune-system and renders it more susceptible to viral infection, as demonstrated by the uneven distribution of cases during the current COVID-19 pandemic. The World Health Organization's recommendations are expressed in terms of air quality index. There is currently no nature-based, efficient and cost-effective solution that can actively create clean urban-air microclimates with sufficiently low AQI values. PhotoSynthetica™ provides localised microclimatic purification precisely where most people live, work and play, thus increasing its effectiveness.

And while new buildings are more energy efficient, only 2% of the UK building stock is less than 5 years old. Therefore, in order to achieve current air quality and carbon neutrality targets there is a need for easily deployable technologies to improve energy efficiency in both new and existing buildings.

Alternative green building technologies such as PVs or solar panels are made of non-renewable and non-recyclable materials, have high embodied CO2, low and rapidly decreasing efficiency and are an eyesore on most building applications.

PhotoSynthetica™ offers a hi-tech nature-based solution. It is fully integrated, customizable and 100% recyclable. The PhotoSynthetica™ protocol guarantees the adaptive behaviour of the system and substantial performance enhancement over time. Its key advantages include adaptive shading and reduction of cooling energy loads, enhanced natural light diffusion, biomass production, on-site bio-energy, and the reduction of electricity usage.

Biomass production represents a unique opportunity to enable the emergence of new circular economies in the urban realm. The increasing global urban population requires substantial amounts of food and products that are currently grown or produced in factories at great human and environmental cost. The global health crisis is a clear manifestation of the unsustainable imbalance of the current system.

New urban circular economies must be developed to support the transformation of waste and pollutants into valuable raw materials and sustainable food production. PhotoSynthetica™ uses solar energy to grow microalgae that contain up to 60% vegetable proteins, enabling fresh food to be grown in an urban environment with little occupation of space. Algal biomass is now also a critical ingredient for emerging sustainable industries in the bioplastic sector, eco-fashion, and in the pharmaceutical, cosmetics and wellness sectors.

Design innovation

The PhotoSynthetica™ system is a design innovation that integrates three core components: wetware, hardware and software.

- The '**wetware**' is living microalgae cultures. The careful selection, integration and management of these powerful photosynthetic organisms guarantees an efficient, resilient and low maintenance system. This is the metabolically active layer of PhotoSynthetica™, converting what the city expels, carbon dioxide, pollutants and other waste, into a promising source of renewable biomass that is currently at the core of several new bio-based circular economies.

- The '**hardware**' is the artificial habitat for cultivation of the living cultures, the so-called photo-bioreactor. It combines digital design and fabrication technologies to optimise spatial and architectural integration in different application scenarios. This makes the PhotoSynthetica™ system suitable for both new and retrofit applications, indoor and outdoor systems.

- The '**software**' is the digital monitoring and management system. It optimises the performance of the living cultures and their ability to adapt to any urban microclimatic condition and to specific patterns of use. The system is connected digitally to sources of big data analysis and uses several sensors to directly monitor the culture's growth conditions and human interactions.

Wetware

Algal biodiversity

Microalgae are unicellular eukaryotic organisms and are among the oldest on Earth. They are predominantly photosynthetic and can be found in a wide range of habitats, particularly marine ecosystems[3]. Microalgae are responsible for a significant portion of the oceans' biogeochemical cycling and constitute the biggest carbon sink on Earth, with more than 50% of the overall CO_2 being absorbed by these organisms.[4]

Microalgae have a varied evolutionary history with genes derived from photosynthetic organisms, heterotrophic eukaryotes and bacteria. This has led to a wide range of adaptations, allowing them to thrive in a variety of conditions. Microalgae in coastal regions are adapted to turbulence, high nutrients and low light whilst open ocean microalgae have to contend with high irradiance and low nutrient concentrations. Polar or alpine

3 Brocks, J., Jarrett, A., Sirantoine, E., et al. (2017). The rise of algae in Cryogenian oceans and the emergence of animals. *Nature* 548, 578–581. doi:10.1038/nature23457

4 Andreeva, N.A., Melnikov V.V., & Snarskaya, D.D. (2020). The role of cyanobacteria in marine ecosystems. *Russian Journal of Marine Biology* 46, 154–165. doi:10.1134/S1063074020030025

microalgae are adapted to freezing temperatures, high nutrients and long periods of light and darkness. This unique diversity, resilience and photosynthetic efficiency makes microalgae particularly suitable for cultivation, not only in natural habitats, but also in urban ones, including dense and polluted cities.

The PhotoSynthetica™ wetware includes a catalogue of microalgae species that are well adapted to artificial habitat cultivation, exhibit intensely different coloration and are capable of re-metabolising pollutants into biomass with a number of valuable applications in emerging green industries.

Here is a list of selected species, grouped by visual appearance, that can be found in many of ecoLogicStudio's recent projects.

■ **Chlorella** is an attractive super-food source because it is high in protein and other essential nutrients. When dried, it is about 45% protein, 20% fat, 20% carbohydrate, 5% fibre, and 10% minerals and vitamins. Mass production methods are now being used to cultivate it in large artificial circular ponds as well as in photo-bioreactors. It is also abundant in calories, fat and vitamins.[5] **Uses:** health food, food supplements, feed surrogates.

■ **Spirulina** is a cyanobacteria. It can be consumed by humans and other animals. It is usually taken by humans as a nutritional supplement and is made primarily from two species of cyanobacteria: *Arthrospira platensis* and *Arthrospira maxima*. Spirulina was a food source for the Aztecs and other Mesoamericans until the 16th century. It has also been traditionally harvested in Chad to make broths for meals. **Uses:** health food, cosmetics.

■ *Tetraselmis suecica* is a green marine alga. It grows as single, motile cells visible under light microscope up to concentrations over one million cells per millilitre. It can be grown as a foodstock in aquaculture, being amenable to species such as rotifers of the genus Brachionus. **Uses:** foodstock.

■ *Chlamydomonas nivalis* is a green microalgae that owes its red colour to a bright red carotenoid pigment, which protects the chloroplast from intense visible and ultraviolet radiation, as well as absorbing heat. Its algal blooms may extend to a depth of 25cm (10 inches), with each cell measuring about 20 to 30 micrometres in diameter. It is widely used as a source of beta-carotene. **Uses:** pharmaceutics, cosmetics.

■ *Phaeodactylum tricornutum* is a blue diatom. It is the only species in the genus Phaeodactylum. Unlike other diatoms *P. tricornutum* can exist in different morphotypes (fusiform, triradiate, oval), and changes in its cell shape can be stimulated by environmental conditions. **Uses:** nutrition, fuel production.

5 Nicoletti, M. (2016) Microalgae nutraceuticals. *Foods* 5 (3), art. 54. doi:10.3390/foods5030054

■ *Haematococcus pluvialis* is a fresh water red microalgae species of Chlorophyta. This species is well known for its high content of the strong antioxidant astaxanthin, which is important in various pharmaceuticals, and cosmetics. The high amount of astaxanthin is present in the resting cells, which are produced and rapidly accumulated when the environmental conditions become unfavourable for normal cell growth. *Haematococcus pluvialis* is usually found in temperate regions around the world. **Uses:** pharmaceutics, cosmetics.

■ *Porphyridium cruentum* is a red marine microalgae and it serves as a model for many laboratory works. The high content of EPA, a long chain essential fatty acid of the omega-3 family makes this alga interesting for the preventive treatment of cardiovascular disease. **Uses:** pharmaceuticals, cosmetics, nutrition.

The careful selection of microalgae species in relationship to site microclimate, urban context and business model is a key asset of the PhotoSynthetica™ protocol. It guarantees resilience, low maintenance and represents a promising source of renewable resources, currently at the core of several new bio-based circular economies.

Photosynthetic potential and carbon sequestration

Photosynthesis is the core process powering PhotoSynthetica™. The process is powered by solar light energy or artificial light energy if it takes place indoors.

Photosynthesis is used by microalgae, PhotoSynthetica™'s wetware, to convert light energy into chemical energy that can later be released to fuel the algae organisms' activities. This chemical energy is stored in carbohydrate molecules, such as sugars, which are synthesised from carbon dioxide and water. Oxygen is also released as a by-product. Here is the overall equation for the type of photosynthesis that occurs in microalgae.

$$6CO_2 \text{ (Carbon dioxide)} + 6H_2O \text{ (Water)} \rightarrow C_6H_{12}O_6 \text{ (Sugar)} + 6O_2 \text{ (Oxygen)}$$

Eukaryotic microalgae and prokaryotic cyanobacteria (both commonly referred to as microalgae) are capable of bio-converting CO2 into microalgae biomass using the electrons released during the light-dependent water photolysis. Most algae and cyanobacteria perform photosynthesis; such organisms are called photoautotrophs.

Photosynthesis is largely responsible for producing and maintaining the oxygen content of the Earth's atmosphere, and supplies most of the energy necessary for life on Earth. PhotoSynthetica™ brings this functionality to the urban realm thus providing new mechanisms to develop carbon neutral cities.

Microalgae are incredibly efficient photosynthetic organisms. We have estimated that compared to large trees PhotoSynthetica™ is 10 times more efficient at capturing carbon dioxide. This is because large trees and other plants are complex multicellular organisms with complex metabolisms supported by a variety of dedicated systems. These systems include roots, trunks, branches, flowers and so on. None of these systems is actively photosynthetic, therefore they cannot participate actively in the CO_2 re-metabolisation process.

Microalgae are simple, single cell organisms, adapted to a symbiotic existence as part of large colonies or as symbiont of other large organisms. Their main purpose is to be actively photosynthetic and they dedicate their entire body mass to the task. As such, when cultivated within PhotoSynthetica™'s photo-bioreactor microalgae can achieve similar carbon sequestration performance to a large tree with 100 times less weight and a minimal occupation of space.

For owners, managers and developers of buildings in both public and private sectors, PhotoSynthetica™ can therefore provide 30% more value compared to green solutions currently on the market. This is achieved through a combination of several advantages that make it a superbly resilient solution. PhotoSynthetica™ creates an adjustable and intelligent solar shading system thus reducing running costs, energy consumption and associated CO_2 emissions. Solar energy is intercepted and captured to power photosynthesis, saving up to 65% of the HVAC cooling energy. Upfront installation costs are 10% less than in a standard vertical garden, and maintenance costs are minimised and can be completely neutralised when the system is fully calibrated such that the value of the harvest offsets them completely. The benefit in biomass production can be significant, with Spirulina currently selling for up to £200/kg.

Installing PhotoSynthetica™ causes negligible loss of leasable space as the bioreactor membrane solution can be as thin as 30cm.

Finally, there is the added value of CO_2 sequestration and air purification as well as increased likelihood of obtaining building performance certifications, which can raise a building's market value by as much as 20%.

Air filtration and pollutants re-metabolisation

We have been running several test bed and pilot schemes from which we have been able to record the exceptional performance of PhotoSynthetica™ in terms of its CO2 absorption.[6]

6 Pasquero, C., Poletto, M., & Greskova, T. (2020). Photosynthetic architecture in times of climate change and other global disruptions. In L. Werner & D. Koering (eds), *Proceedings of the 38th eCAADe Conference*, vol. 1. Berlin: eCAADe.

Microalgae are also known to be able to absorb most particles, including metals and gasses, once they are exposed to them in their aquatic medium. Furthermore, their accelerated metabolism and growth rate make them very effective at this filtering task. For this reason microalgae are already commonly used a part of biological wastewater filtration and purification processes. Lab tests have demonstrated that microalgae have unrivalled abilities to capture and re-metabolise urban air pollutants too, especially NOx and SOx. In urban environments, these properties combine with our proprietary hardware design as well as monitoring software to achieve significant reduction in local concentrations of pollutants. In some occasions, this combined effect can prevent the formation of ground level ozone, another harmful component of urban smog. The AirBubble project, developed in collaboration with GSK, is our pioneering development in this context.

Scientific literature confirms the potential of microalgae and their suitability for this task. Using microalgae to clean up flue gasses via photosynthesis is considered a promising CO2 mitigation process,[7] and algal growth by assimilation of NOx is claimed to promote NOx removal rates from the gas phase by up to 97%.

7 Yen, H.-W., Ho, S.-H., Chen, C.-Y., & Chang, J.-S. (2015). CO2, NOx and SOx removal from flue gas via microalgae cultivation: A critical review. *Biotechnology Journal* 10, 829–839. doi:10.1002/biot.201400707

In short, algae are capable of using the carbon and nitrogen molecules as food to grow their biomass thus removing them from the air and trapping them in their cells. PhotoSynthetica™ hardware technology optimises this process by increasing the surface area of contact between algae and gas. It also optimises the surface area of contact between the photo-bioreactor's surface and air pollution at the urban scale. In other words PhotoSynthetica™ effectively deploys microalgae's metabolic power to transform the urban microclimate and related local air quality.

Our AirBubble project has confirmed the unique potential of this biotechnological approach to microclimatic urban filtering, and more field tests are under way to gather comprehensive data and accelerate the optimisation protocol, thus increasing the efficiency of the technology.

Crucially, since PhotoSynthetica™ is a living bio-digital system, it can be 'trained' and it will adapt in time to specific site conditions, progressively improving its performance through use.

Widespread use will therefore demonstrate its benefits for human health in multiple scenarios. This is achieved not only because of the algae properties but also and perhaps foremost because of the strategic design of the PhotoSynthetica™ system and its architectural integration and material customisation to unique site conditions. This unique level of wetware, software and hardware customisation enables PhotoSynthetica™ to grow algae where it is most effective for screening people from harmful air pollution.

In conclusion, while carbon sequestration is a process that can be dislocated spatially, filtering air pollution is only really effective if it operates strategically where emitters are located and where people are most likely to inhale the harmful particulates and gasses.

Growth protocol

The efficiency of PhotoSynthetica™ depends on the ability of the living cultures to grow consistently in any urban condition. The proprietary, software-monitored and AI-integrated, growth protocol makes this possible.

Microalgae are aquatic organisms and therefore necessitate a growth medium to thrive. Typically, this is made of purified water plus a combination of nutrients. PhotoSynthetica™ also includes a jellified version for special applications. The conditions of the medium must be monitored carefully to make sure it supports the growth of the microalgae. Several parameters are monitored, including temperature, pH, salinity, conductivity and oxygen levels.

For each species of algae, there are specific ranges of tolerance for each parameter. These ranges have been first set in the lab and, for some species, were further tested in large-scale prototypes. These tests inform the growth protocol's default settings. These settings are continuously monitored and adjusted once the system is in operation leading to constant self-improvement.

The natural biological adaptation of the cultures to their environment and self-improvement of the growth protocol combine to give PhotoSynthetica™ exceptional adaptive efficiency. In other words, whereas all traditional green technologies, such as wind turbines or solar panels lose efficiency in time and become obsolete after about 20 years, PhotoSynthetica™'s performance will improve over time and with future upgrades.

The proprietary growth protocol has been developed in collaboration with microbiologists as well as algae farmers and has been tested in different concentrations and monitored in indoor and outdoor conditions. To go a little more into the details of its structure, what follows here is a high-level description of the case of Spirulina *(Arthrospira platensis)*.

- **Light.** Spirulina is a photosynthetic bacteria (a blue-green algae is actually a bacteria). This means that just like a plant, Spirulina needs the energy from the sun to live. It can be exposed to a maximum of natural sunlight to boost production, but this should never exceed 120,000 lux. In extreme light conditions Spirulina may bleach within minutes and eventually die. In this case Spirulina cells become 'lysed' (cellular lysis is the breakdown of a cell and compromises its integrity) and when that happens light intensity must be reduced immediately to avoid losing the entire culture.

If the culture appears yellowish or green-olive the Spirulina cells are under photo stress (too much light) and the cells are lysing—there is too much light for the chlorophyll to deal with and the cultures must be shaded.

If the culture is yellowish and foaming, it means that lysis has already occurred and that the cells have broken open. The foam is the polysaccharides released from the cells into the culture medium. In this case shading must be combined with an increase in the agitation of the medium, lowering of the pH and addition of nitrogen and potassium.

There are four specific circumstances when shading is critical: if the culture is very warm (over 37°C), or it is very cold (below 14°C); if the culture is recovering from any problem and is still struggling; if the culture is very diluted and the Spirulina cells are not self-shading; and when the available natural light is not sufficient—artificial wide spectrum lamps can then be used to boost growth. Spirulina has a unique set of pigments (called phycocyanins and allophycocyanins) in addition to the chlorophyll, which allow it to absorb more red and orange light wavelengths than blue and green wavelengths. For this reason, it is best to use lights that have a warmer colour (625–650 nm). If artificial light is required to completely replace natural light it should be timed to a 16-hour day cycle to allow for a period of overnight resting.

- **Temperature.** Temperature directly influences the growth rate of the cultures fairly drastically. Spirulina can survive cold temperatures down to about 3–5°C. It will begin to grow when the temperature is 14–15°C and growth becomes increasingly visible over 18°C. At 18°C the growth rate is only about 50% of the growth rate at 20°. At 20°C the growth rate is only 55% of the growth speed at 22°C. And between 22°C and 32°C the grow rate doubles. However, between 32° to 35°C the growth rate peaks, and above 37°C Spirulina risks becoming damaged. After several hours above 44°C the cultures will almost certainly be killed. Therefore, the optimum temperature for maximal growth is between 32°C and 37°C. Moreover, if the climatic fluctuations of temperature are wide, it becomes necessary to add thermal mass or insulation since the cultures may easily become stressed.

■ **Aeration.** Spirulina is a photoautotrophic organism, it 'makes' or 'builds' its substances by photosynthesising all the nutrients and minerals that are dissolved in its environment (i.e. water). In order to do that effectively, Spirulina uses the energy from the sun that is gathered by its different pigments. All photosynthetic cells have upper and lower limits to the amount of sun they can be exposed to. Agitation (stirring) via medium recirculation or bubbling is the best method for achieving optimal exposure to all of the cells floating in a given volume of medium. Multiple systems have been tested to provide CO_2 bubbling to the algae using air from the urban realm or from dedicated sources. These vary from the manually activated, to the mechanically operated, to the fully electronically controlled. Manual systems include small manual pumps as well as larger foot pumps. The mechanically operated systems consists of air pumps for the bubbling and/or water pumps for the recirculation. Air flows have been tested from as little as 20 litres per minute to as large as 2000 litres per minute.

■ **Culture medium.** The culture medium is of the utmost importance for Spirulina growth as the algae cells will absorb whatever is in their environment. This means that they will bind with heavy metals and other toxins as well if present in the liquid medium. If Spirulina is cultivated for food purposes it needs to be grown in purified water using a carbon filter. Once Spirulina cultures are dense and thriving there is very little risk of contamination mainly due to the growing conditions that are very specific to Spirulina (pH over 9.6 and a Secchi lower than 3). Spirulina has basic needs that include: an alkaline water, a source of fixed nitrogen, a source of iron, phosphorus, potassium and sulphur, a source of carbon (this can be the CO_2 atmospheric) and a source of trace calcium, chlorine and magnesium.

Among the many available medium recipes the 'Zarrouk' medium is very effective and supports a very good Spirulina growth rate.

This bio-gel medium has been tested in laboratory conditions and monitored with a microscope in collaboration with the Synthetic Landscape Lab at the University of Innsbruck and has also later been tested in large-scale temporary installations.

■ **Harvesting and maintenance protocol.** When the culture is dense enough, with a Secchi reading of 3 cm, it is possible to harvest biomass from it. Typically, one third of the culture's volume can be harvested at any given time. The operation can be repeated when the overall density of the culture has again reached the minimum recommended level. The separation of the biomass from its growth medium can be achieved with various processes and techniques leading to different degrees of harvesting efficiency.

Circular economies

One of the major advantages of the PhotoSynthetica™ technology is that biomass, once harvested, can be used in a variety of ways. Therefore, the maintenance costs associated with running the system can be paid for by the value of the biomass extracted. On larger and more sophisticated applications, where the growth protocol is automated and the extracted biomass is of high quality, earnings can be significantly larger than costs, guaranteeing a good return on investment in the system.

Early adopters of PhotoSynthetica™ have two main options when it comes to starting new circular economies based on the use of biomass grown on-site: either managing the system internally and profiting directly from the biomass extracted, or subcontracting maintenance to one of PhotoSynthetica™'s local algae-farming partners, thus contributing to the emergence of a distributed business ecosystem.

Several emerging industries are already utilising algal biomass as part of their supply chain, so the adoption of PhotoSynthetica™ can enable the creation of new sustainable business ecosystems and new supply chain synergies that lead to carbon neutral material cycles. Theses are some of the most relevant possibilities:

■ **Bioplastic.** The bioplastics or organic biodegradable plastics industry extracts value from renewable biomass sources such as vegetable oil and starch that unlike components of fossil-fuel plastics are not derived from petroleum. Bioplastics provide the twin advantages of conservation of fossil resources and reduction in CO_2 emissions, which make them an important innovation in sustainable material development. Packaging, eco-fibres and eco-foams for the fashion industry are key applications for bioplastics. Algae serve as an excellent feedstock for plastic production due to their many advantages such as high yield and the ability to grow in a wide range of environments. Algae bioplastics mainly evolved as a by-product of algae biofuel production, where companies were exploring alternative sources of revenues along with those from biofuels. ecoLogicStudio has adopted bioplastic membranes and 3D-printing bioplastic technologies in several PhotoSynthetica™ hardware components. We are also currently developing a line of algae bioplastic 3D-printing filaments, as well as carbon neutral algae bioplastic products and packaging for the pharmaceutical and food industry.

■ **Nutraceutics and future food.** Microalgae's chemical composition is a complex mixture of minerals, vitamins, and primary and secondary products, offering a large spectrum of possible applications and utilisations for humans and animals alike.

The nutritional value of Spirulina, for example, was already known to the Aztecs, who harvested the alga from Texcoco Lake, near Mexico City. Spanish soldiers led by Cortes described its use as a daily food and the sale as cakes. It is rich in vitamins, minerals, β-carotene, essential fatty acids, and antioxidants, all of which have facilitated its commercial production as a human food supplement over the course of the past decade.Its consumption has been shown to have cardiovascular positive effects, lowering blood pressure and reducing cholesterol. In consideration of its anti-carcinogenic properties, it was used to treat radiation sickness in people that were affected by the 1986 Chernobyl nuclear accident. ecoLogicStudio is currently developing a bio-factory concept for the food industry integrating the PhotoSynthetica™ technology on the walls of the main food production facilities.

■ **Cosmetics.** Extracts from algae are proving extremely useful in anti-aging skin care by providing improvement to the firmness and elasticity of the skin. Extracts from algae also exhibit retinol-like properties, antioxidant properties, tensioning properties, cell renewal enhancing properties and hydrating properties. The availability of fresh algae in the urban realm enabled by the adoption of PhotoSynthetica™ will greatly enhance the power and cost effectiveness of biocosmetics and open up new applications that necessitate the use of fresh algae.

Hardware

The hardware of PhotoSynthetica™ provides the artificial habitat for cultivation of living cultures. Its core component is the photo-bioreactor, where algae are exposed to the sun and where photosynthesis takes place. PhotoSynthetica™ combines digital design and fabrication technologies to optimise the photo-bioreactor's architectural integration, environmental performance, aesthetic quality and customisation.

Photo-bioreactor technology

The PhotoSynthetica™ system includes three kinds of photo-bioreactor technology, maximising the versatility of the system for urban applications in both new and retrofitted building scenarios, as well as indoor and outdoor conditions.

- **Membrane technology.** There are two kinds of PhotoSynthetica™ membrane photo-bioreactors. Permanent reactors are made of ethylene tetrafluoroethylene (ETFE) or aliphatic thermoplastic polyurethane (TPU) while temporary applications are made of bioplastic or biodegradable TPU. Permanent photo-bioreactors employ a double ETFE foil of 200 microns thickness each, with a gap for algal flow of max 3 cm in the thickest point. The double ETFE foil is welded together with a proprietary pattern that guarantees optimal algae flow, reactor robustness and minimises deformability. Within clearly defined design limits each reactor can be uniquely shaped as part of an overall building cladding pattern. Reactor membranes are edged with rubber profiles and clamped with aluminium profiles along the main edges and bolted to the supporting building structure. During fabrication, the two separate layers of ETFE are welded following digitally designed patterns derived from an accurate 3D model of the final installation, a guarantee of absolute precision even in cases of building retrofit. Welding tools used are precision machines based on the impulse welding technology and handle a maximum welding temperature of 400° Celsius.

The advanced mechanical and chemical strengths of ETFE gives PhotoSynthetica™'s bioreactors unrivalled transparency (superior to glass), durability (around 20 years), lightness, total recyclability and excellent fire safety rating, ideal for large-scale permanent architectural applications.

When applications are temporary or a more economical solution is sought, such as for instance the wrapping of temporary scaffolding or large warehouse buildings, ETFE is substituted by PolyAir bioplastic foils or TPU membranes. PolyAir is a starch-based bioplastic foil that at the end of its useful lifetime can be fully recycled. This material offers sufficient mechanical strength, translucency and 3 years' durability. Most importantly, it is a carbon neutral material, which, combined with PhotoSynthetica™'s carbon capturing capabilities, makes the final composite solution carbon negative.

■ **Glass technology.** For indoor as well as special built outdoor applications (courtyards, large lobbies, public space designs), laboratory grade glass reactor technology has been developed. Based on the Schott borosilicate glass tubes, PhotoSynthetica™ customises the fabrication and connection process to adapt to a variety of architectural spaces. The process starts with a full, 3D digital model of the complete photo-bioreactor and of the supporting structure developed in the Rhinoceros CAD environment and utilising a custom code. Engineering and detailed design are also completed in a full 3D environment.

A new virtual reality output is then produced to enable the model to be checked and shared with fabricators and stakeholders. The full model is then unfolded in 2D shop drawings and fabrication files. Connecting elements are 3D-printed in polycarbonate or polylactic acid (PLA) bioplastic using fuse filament technology with Ultimaker 3 and WASP 40/70 industrial machines. A full assembly manual and procedure are issued for each custom application.

The glass technology allows for unrivalled aesthetics, fire resistance, durability and easy maintenance. Cultures also seem to thrive in glass tubular reactors, making this the ideal solution for high end, high productivity architectural systems for the food, nutraceutical and pharmaceutical industries.

■ **3D-printing technology.** The most technologically advanced solution to date are 3D printed photo-bioreactors. While the solution is still at an advanced experimental stage, progress is fast due to the incredible evolution of current 3D-printing technology. Large-scale 3D printing is now affordable and offers incredible precision of execution, high resolution, material efficiency, full recyclability or biodegradability and a waste free fabrication process.

3D-printing technology also enables a higher degree of articulation of the large-scale photo-bioreactors, and this is particularly interesting for solutions involving air purification and filtering. In the PhotoSynthetica™ 3D-printed bioreactors, the supporting material is deposited at a resolution that allows the formation of triangular bio-pixels. The composite solution works like a porous sponge that allows air to circulate through it.

The final aesthetic qualities are also unrivalled, making this solution ideal for high-end indoor and outdoor environments and prestigious settings.

Once the fully 3D-engineered model is produced it is then exported to the CURA software for slicing. Machine code is exported. The 3D-printing process uses the fuse filament technology, polycarbonate or PLA bioplastic materials and currently takes place on a set of Ultimaker 3 and WASP 40/70 industrial machines. The layering process is algorithmically controlled driving the actual tool paths of the 3D-printing nozzle. In this way, the digital description is perfectly translated into lines of deposited material. Each deposited layer is 400 microns thick with triangular infill units of 46 mm. Full assembly manual is issued.

In 2021 a pioneering collaboration with Swarovski has led to the world's first 3D-printed glass reactors and glassware set, presented as part of the project Bit.Bio.Bot at the Venice Architecture Biennale.

Circulation and harvesting systems

There are two kinds of systems to manage the algae cultures within the photo-bioreactors in PhotoSynthetica™. A closed loop circulation system (with or without aeration) and a simple aeration system.

■ **Closed loop circulation.** In the closed loop system, the cultures are kept in a constant flow. The loop starts in the cultures tank, in glass, acrylic or polypropylene. The cultures are then pumped by an immersion pump to reach the highest point in the system. From here they flow down by gravity through the photo-bioreactors where they expand, meet with incoming light, photosynthesise, emit oxygen and then mix into a turbulent motion. Once the cultures reach the tanks again they are allowed to fall freely, thus releasing the oxygen in the atmosphere. The tank is aerated with an air pump or compressor and CO_2 is introduced to the culture before a new loop starts. This system of circulation is optimised with valves to operate in connection with membrane photo-bioreactors.

A more sophisticated use of aeration has been implemented with glass tube photo-bioreactors in projects such as the Bio.Tech Hut. This solution uses high velocity air flow to lift the living medium up into the lab-grade glass tubes. The fast air stream creates eddies on the liquid surface, and a unique stirring effect that gives the desired O_2–CO_2 exchange. The bubbles that form during this process keep the wall of the tubes clear, improving the quality and productivity of the cultures.

The fluid then descends by gravity to complete the loop. Multiple loops can be coiled around inhabited areas thus increasing the cultivated volume without loss of usable space. This system is extremely intensive and efficient in the use of space and energy. The Bio.Tech Hut project is a pioneering application of this technology.

■ **Simple aeration.** In this simple aeration system the cultures are contained in the photo-bioreactors and are bubbled or aerated in order to be maintained in constant motion. Air, either clean or unfiltered urban air, is introduced at the bottom of the reactor. As the air bubbles rise through the watery medium within the photo-bioreactor they come into contact with the voracious algae cells. CO_2 molecules and air pollutants are captured and stored by the algae and grow into biomass. Freshly photosynthesised oxygen is released at the top of each reactor unit. Reactor design often includes a serpentine pattern to optimise the carbon sequestration process by increasing the surface area and time of contact between algae cells and air molecules. This system is more simple and cost effective and can still lead to excellent yields.

An experimental evolution of this system is under testing and involves a completely passive aeration model through a porous 3D-printed reactor. In this system algae grow on a jellified medium and are exposed to incoming wind flow. The system is demonstrating excellent air filtration properties but does not currently incorporate a harvesting system.

■ **Harvesting systems.** PhotoSynthetica™'s harvesting system includes three options.

Micro-screen, micro-strainer and micro-cloths are the simplest and cost effective techniques suitable for smaller and low-tech applications or for enhanced user interaction. The BioBombola is a beautiful example of this logic at work. Its function is principally based on passing or retaining cells that are introduced onto a sieve of a given aperture size. These techniques are limited in efficiency and can only be adopted with microalgae of large cell size, such as Spirulina, with a separation size larger than 70 μm.

Centrifugal machines instead work using the sedimentation principle, where the centripetal acceleration causes denser substances to separate out along the radial direction (the bottom of the tube). Optimisation requires knowledge of algae properties such as algal size and oil content. Centrifuge machines are expensive and only become cost effective for larger systems.

Finally, flocculation is used to describe the removal of a sediment from a fluid. In addition to occurring naturally, flocculation can also be forced through agitation or the addition of flocculating agents. Typically a flocculant has a different electrical charge than the substances being precipitated out of the solution, attracting the material to itself in clumps of particles. Clumps can then be easily filtered out of the culture medium.

Monitoring systems

The real-time monitoring system of PhotoSynthetica™ includes a series of sensors that can be installed independently. The core sensors are: aquatic sensors, environmental sensors, proximity and occupancy sensors, visual and camera sensors.

■ **Aquatic sensors.** These sensors monitor algae growth conditions inside the liquid medium. The sensors typically measure the four main factors affecting algal growth and metabolism such as salinity, pH, oxidation reduction potential and temperature. The system has Wi-Fi built in so it can update the monitoring controllers and enable automation of key farming actuators. The aquatic sensor monitors data every 10 minutes. It currently works as a cloud-based service on nearly every device platform. Once the thresholds for the sensor are set up, it can send notifications and alerts on things such as over/under temperature, pH level and salinity level.

■ **Environmental sensors.** To register microclimatic conditions around the photo-bioreactors environmental sensors are deployed. These are typically temperature, humidity and light sensors. Their input is used to automate key actuating devices to moderate climatic extremes that may otherwise damage the cultures.

■ **Occupancy sensors** are used to register human presence and patterns of use. Typically, these are proximity sensors and accelerometers, measuring distance and speed of movement. These parameters affect the actuation of the system to guarantee optimal human comfort by providing adaptive shading and glare reduction.

■ **Visual and camera sensors** use webcam technology to monitor the qualitative changes in the cultures, with special focus on colour mapping and detecting early signs of stress in the cultures. These are linked to warning messages in the farming protocol requesting direct human intervention. These sensors are currently under further development to integrate machine vision into the gardening protocol, leading to a new AI 'virtual gardener' interface.

Actuating systems

Actuating systems implement a more advanced degree of automation, interaction and self-regulation in the gardening protocol. The actuators tested so far include solenoid valves for algae flow regulation, a mister system for microclimatic evaporative cooling, radiant heaters for culture warming, QR codes for rapid data access and system augmentation, and wide spectrum lighting for enhanced growth.

Actuators are controlled automatically and respond to the growth protocol's instructions in real time.

Software

The wetware is the core active ingredient of PhotoSynthetica™, while the hardware provides the ideal artificial habitat for its growth. However, in order for the wetware to perform efficiently and evolve robustness in such a complex and fast-changing environment as the urban realm, it is necessary to augment the biological living cultures with digital sensibility and computational intelligence.

PhotoSynthetica™ software includes a package of algorithms and interfaces connected with embedded sensing and actuating mechanisms. It is able to record parameters from the micro algae culture medium and from the surrounding urban environment in real time, and inform design choices, patterns of use, manufacturing and maintenance protocols.

Furthermore, a new interface for biologically augmented artificial intelligence (BAI) is under development. The ambition of this upgrade is to achieve overall self-regulation and self-sufficiency in the system and in its interaction with human users. Ultimately, the system will require little regular maintenance and it will evolve bio-autonomy. PhotoSynthetica™ has established a collaborative and growing ecosystem of researchers and partner companies to help us develop and bring to fruition our innovative software technology.

Microclimatic analysis

Knowledge of the specific local urban environment is critical for the PhotoSynthetica™ system and its living cultures. This is true during both the detailed design and operational phases. The critical factors are light, temperature and aeration.

- **Light.** Access to light energy from either the sun or an artificial source is crucial for the system to perform. However, too much light can damage the cultures. In addition, different algae strands have different sensitivities to light. It is therefore critical to simulate and adjust the site conditions for each urban scenario before progressing with the specific application and installation of the system. This knowledge will inform the choice of photo-bioreactors, location and architectural integration, as well as the choice of particular algae species.

PhotoSynthetica™ adopts a 3D environmental modelling approach enabled by the software plug-in Ladybug and the CAD environment Rhinoceros. GIS data from Open Street Maps or a similar geographical database is exported and loaded to develop a highly accurate urban model of the surroundings. For interiors, if survey data is not available, a 3D laser beam is used to scan the environment and to capture details of the dimensions and shapes from the surfaces of the surrounding structures.

Microclimatic data files of the specific city or area are loaded and analysed in Ladybug Grasshopper plug-in. This tool can be used to customise solar analysis and shadow studies. The data is extracted from continuously updated libraries and can also focus on more accurate analysis of lighting levels that is performed on the surface of the installation. This takes into account statistical cloud cover and overshadowing from surrounding constructions or other obstacles. Seasonal and daily fluctuations, including peak conditions, are also simulated. Visual maps and diagrams are generated to inform detailed design decisions.

- **Temperature.** The same 3D environmental modelling approach is also adopted to simulate thermal regimes. Accurate incident solar radiation analysis is performed on the surface of the installation. This takes into account statistical cloud cover and overshadowing from surrounding constructions or other obstacles. Seasonal and daily temperature fluctuations, as well as peak conditions, are simulated. The incident solar radiation is measured in kilowatt hours per square metre (KWh/m^2). This analysis also provides yearly cumulative incident solar radiation ($KWh/y*m^2$). Visual maps and diagrams are generated to inform detailed design decisions.

Urban air quality monitoring

As there are two main aeration protocols for the living cultures in PhotoSynthetica™, the cultures flow naturally or under pressure and create eddies and little waterfalls. Air is introduced and bubbles rise naturally through the liquid medium. If the incoming air is polluted urban air, filled with polluting particles and gasses, PhotoSynthetica™ can function as a biotechnological urban air filter and reduce their concentration in the air.

This is possible due to the unique ability of microalgae organisms. As mentioned earlier, during photosynthesis microalgae grow utilising the sun's radiant energy and this enables them to consume atmospheric CO_2 as well as other gasses. Fine particulate matter is also 'washed' from the air during this process. To optimise this filtering performance and its benefits for human health it is essential to gain knowledge of the pollutants that are present in the air and their relative concentration in the specific installation site.

Typically, there are six urban pollutants that weather stations around the world monitor based on the World's Air Pollution Real-time Air Quality Index. These are: PM2.5, or fine particulate matter smaller than 2.5 microns; PM10 or fine particulate matter smaller than 10 microns; O_3 or ground level ozone; NO_2, nitrogen dioxide; SO_2, sulphur dioxide; and CO, carbon monoxide.

All of these have an adverse impact on human health, which depends on their concentration and time of protracted exposure.

- **PM2.5 and PM10** particles can get deep into our lungs and some may even get into the bloodstream. Exposure to such particles can affect both lungs and heart, bringing premature death, aggravated asthma, decreased lung function, increased respiratory symptoms, such as irritation of the airways, coughing or difficulty breathing.

- **O_3**. Breathing ozone can trigger a variety of health problems including chest pain, coughing, throat irritation and airway inflammation.

- **NO_2** exposures over short periods can aggravate respiratory diseases, particularly asthma. Longer exposures to elevated concentrations of NO_2 may contribute to the development of asthma and potentially increase susceptibility to respiratory infections from virus.

- **SO_x** can react with other compounds in the atmosphere to form small particles. These particles contribute to particulate matter (PM) pollution.

- **CO**. Breathing air with a high concentration of CO reduces the amount of oxygen that can be transported in the bloodstream to critical organs like the heart and brain.

PhotoSynthetica™ adopts a big data approach to air pollution monitoring powered by a custom algorithm and a dedicated graphic user interface (GUI). Urban pollutants data are collected from the weather stations within the specific city or project area, showing their air quality index (AQI). The measurements are then organised on a circular diagram with 365 values (365 days) deviated equally in the main circle and five homocentric circular zones that represent the AQI levels. This protocol then follows a two level analytical approach.

In the first analysis, the average daily pollution concentration data are considered for each of the six key pollutants for a period of one year. The daily AQI is based on the 24 hours average of hourly readings. For each pollutant, the average daily AQI index is calculated and represented in a circular diagram. The data are then reprocessed based on the WHO AQI index system. The target is to reduce the AQI levels to the green zone where concentrations of pollutants are not harmful to human health. Pollutants whose concentrations are hazardous are highlighted. The values of their AQI are typically plotted in the red or purple zones.

In the second analysis, pollution concentration data are considered for each of the six key pollutants and a six years comparison is drafted. For each pollutant the average daily AQI index is plotted on a circular diagram of one-year data, for a total of six diagrams. The data are then reprocessed based on the WHO AQI index system. The target is to reduce the AQI levels to the green zone where concentrations of pollutants are not harmful to human health. The final combined visualisation across the six year period is then created.

Growth optimisation

The light, radiation and air quality analysis of PhotoSynthetica™ combine into an advanced urban environmental model and guarantee optimal design and integration of the system in different urban scenarios. The actual resilience of the system and its ability to self-regulate however rely primarily on the integration of this simulation model with real-time data monitoring, enabled by the sensors embedded in the system itself.

The aquatic sensor monitors algae growth conditions within the photo-bioreactors. Daily data are typically captured every 10 minutes. They are then plotted on a circular diagram with 144 values. The minimum/maximum values for each calibration of each culture are represented on the diagrams with homocentric coloured circles. For example, in the case of a Spirulina culture, oxygen reduction potential (ORP), pH value, salinity and temperature are monitored within the liquid medium.

If the ORP sensor returns a value less than 200mV then oxygen bubbling needs to be activated and/or the pH decreased. If it is more than 450mV, the oxygen needs to be reduced and/or the pH increased. This kind of real time adjustment makes it possible for the system to self-regulate and in time become more robust against drastic microclimatic fluctuations or patterns of use.

Similarly, if the pH is recorded over 11, then the medium needs to be renewed and the strain is harvested. Spirulina's growth rate is quickest at around pH 9.6.

The culture's appearance gives significant feedback on its health. Colour changes as well as foaming and sedimentation patterns on the surface of the reactor are early signs of stress. In other words, the cultures act as biosensors toward changes in the surrounding environment. Our ability to read and interpret those signs is significant in two ways: it guarantees the robustness of the system itself and also enables us to be aware of the changes in the urban environment we inhabit. Recent progress in AI design techniques, pioneered by ecoLogicStudio in projects like the GAN-Physarum and machine vision (under current development), are combined to create a new level of embedded intelligence in the PhotoSynthetica™ system. The final objective is to achieve complete bio-digital responsiveness towards a constantly evolving system performance.

This is essential to validate simulated data and to inform the aeration, harvest and maintenance protocols. Managing the culture's flow rate, aeration, bubbling and overall density through programmed harvests is the best way to optimise the system's resilience and achieve the goal of a living, self-sustaining bio-digital architecture.

Air pollution filtering

The air pollution filtering performance is tested in three phases: static local monitoring, geo-located local monitoring and comparative analysis. The first phase establishes local ambient air quality in relation to health-impact thresholds. The second tracks air quality changes within the local microclimate and in the immediate vicinity of PhotoSynthetica™. The third compares these results with the average data from a local weather station and the status of the living cultures in the reactors. The most advanced implementation of this protocol can be found in the AirBubble playground project.

The portable sensor is typically a personal air quality tracker, most recently the Flow2 from Plum Lab. This kind of sensor is relatively accurate and is self-calibrating. Its detection ranges between 0 and 2000 parts per billion or 0 and 200 micrograms. It measures the urban pollutants of NO2(ppb), VOC(ppb), PM2.5 ($\mu g/m^3$), PM10 ($\mu g/m^3$) both indoor and outdoor, based on the local concentration that can be detected, and shows real-time AQI data. Data is captured every minute, processed and plotted on a circular diagram with 1440 values (24h = 1440min). Thresholds correspond to different health risks and are visualised as homocentric coloured circles. This sensor can store up to one month's monitoring data of date, time, pollutants' concentrations as well as the AQI index data.

Portable sensors can also be geolocated, plotting air quality at specific geographical coordinates in the urban realm. The combination of pollution values and location data enables a more granular mapping of the microclimatic nature of urban air pollution, where recorded concentrations can change radically even just a few metres apart. This technique was critical in developing ecoLogicStudio's vision of an air-purifying biotechnological playground. Pollution concentration datasets are directly mapped onto a GIS data plan derived from Open Street Maps to develop an accurate urban model of the specific project area.

Once a more granular description of urban air quality in space and time for a given project site has been produced, it can be compared with historical datasets from the local weather station. The data are compared on a single consistent diagram in order to help tracing correlations in time, predict trends and influence behavioural change.

Conclusions

PhotoSynthetica™ is an advanced nature-based solution to achieve carbon neutrality, purify the urban atmosphere and create new, circular economies. The PhotoSynthetica™ system integrates these objectives in the spaces we inhabit thus enabling a participatory model of urban sustainability. In order to test and develop this model further ecoLogicStudio has designed and built several architectural scenarios and pilot schemes, ranging in scale from regional masterplans to domestic gardening kits.

The following pages provide a comprehensive overview of the most significant case studies arranged by scale, from the small to the very large. Overall, they demonstrate the potential of the PhotoSynthetica™ system and exemplify the scope of the ethical and aesthetical missions of ecoLogicStudio's practice. Our current focus is instigating future pilot schemes with increasing ambition and positive impact.

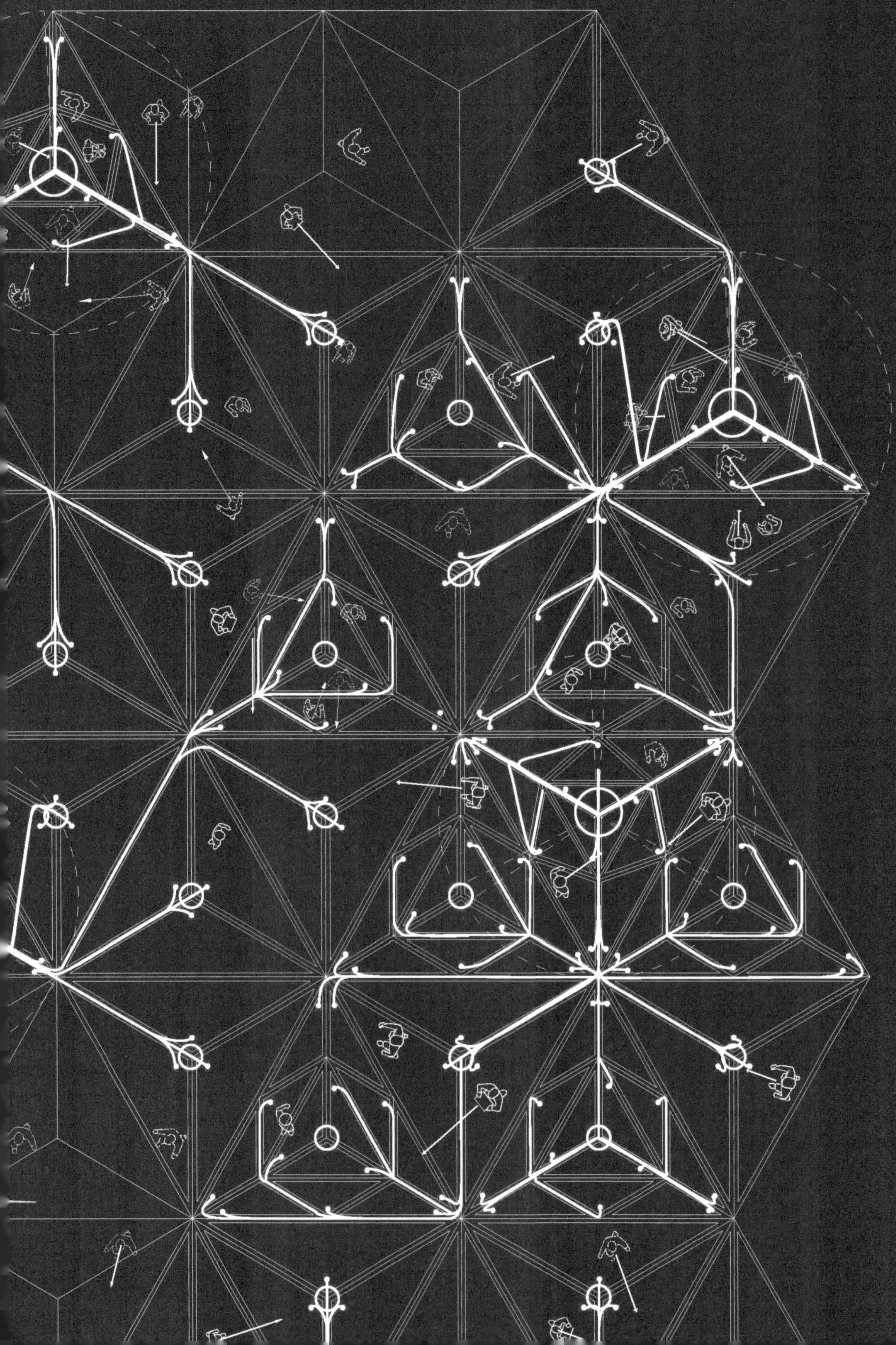

Projects 1.01

PHOTO-SYNTHETICA TEST BEDS, PILOT PROJECTS AND EARLY ADOPTIONS

1.01.01 Domestic realm
BioBombola 40

1.01.02 Cultural and community realms 01
BIT.BIO.BOT 46

1.01.03 Cultural and community realms 02
Storytelling Bio.Curtain 58

1.01.04 Retail and showroom realms
Bio.Tech Hut 64

1.01.05 Architectural retrofitting 01
PhotoSynthetica Dublin 76

1.01.06 Architectural retrofitting 02
PhotoSynthetica Helsinki 84

1.01.07 Architectural retrofitting 03
BioFactory 88

1.01.08 The public realm
Airbubble 98

With BioBombola we invite individuals, families and communities to cultivate a domestic algae garden.

1.01.01

Domestic realm
BioBombola

During the first COVID-19 lockdown in the spring of 2020, the authors bicycled everyday with their two children, Giacomo and Lulu, between their home in Broadway Market and their bio-lab in Hackney Wick (in London's East End) while the children were engaged in a home-schooling programme. The authors decided to involve the whole family in the algae cultivation and in the collection of data on air pollution. The children demonstrated their curiosity in harvesting Spirulina and in baking protein bread with it every week.

Following these spontaneous lockdown-inspired experiments, ecoLogicStudio decided to develop the concept further and to create a minimalist kit for the indoor cultivation of algae. And so BioBombola was born: a pioneering product that invites individuals, families and communities to cultivate a domestic algae garden and sustainable source of vegetable proteins. BioBombola absorbs carbon dioxide and oxygenates in homes more effectively than common domestic plants while also fostering fulfilling daily interaction with nature.

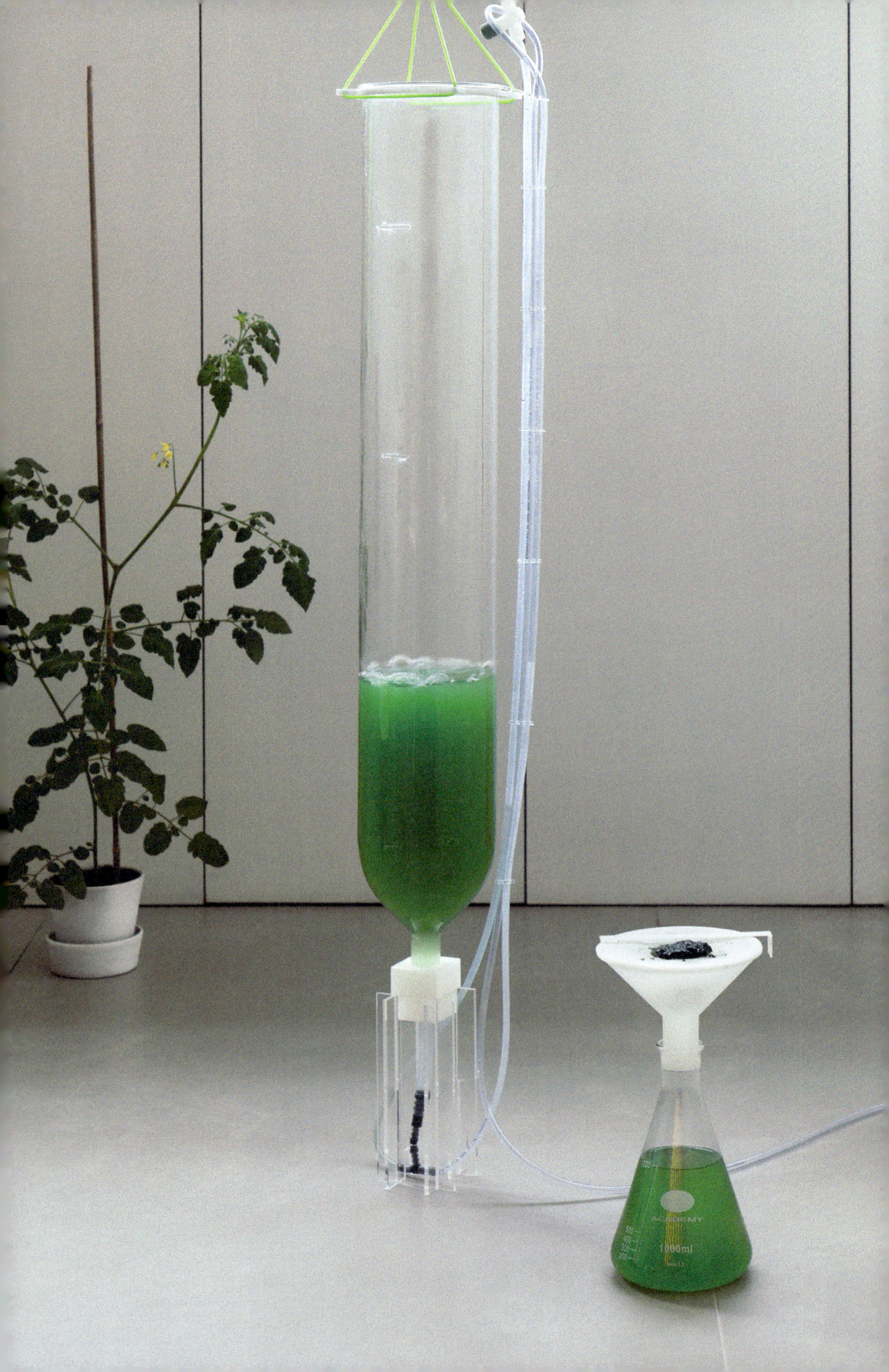
ACADEMY
1000ml

Each BioBombola is composed of a single customised photo-bioreactor, a one-metre-tall, lab-grade glass container filled with 15 litres of living photosynthetic Spirulina strain (a type of cyanobacteria, which is a family of single-celled microbes that are often referred to as 'blue-green algae') and culture medium with nutrients. It also includes an air piping system and a small air pump that constantly stirs the medium. The gentle bubbling from the air system keeps the algae afloat, aids oxygenation and produces a calming sound that emanates with the fresh oxygen in the surrounding environment.

The photo-bioreactor has the equivalent absorption capacity of two young trees in CO_2, while producing the same amount of oxygen as seven indoor plants. The harvest is a simple and entertaining process that can be performed several times per week, collecting up to seven grams of Spirulina per day (one tablespoon), which is the daily recommended intake for a family of four.

Cover page: BioBombola in use in a domestic environment.

This page: BioBombola with the domestic harvesting kit showing details of wet algae growth.

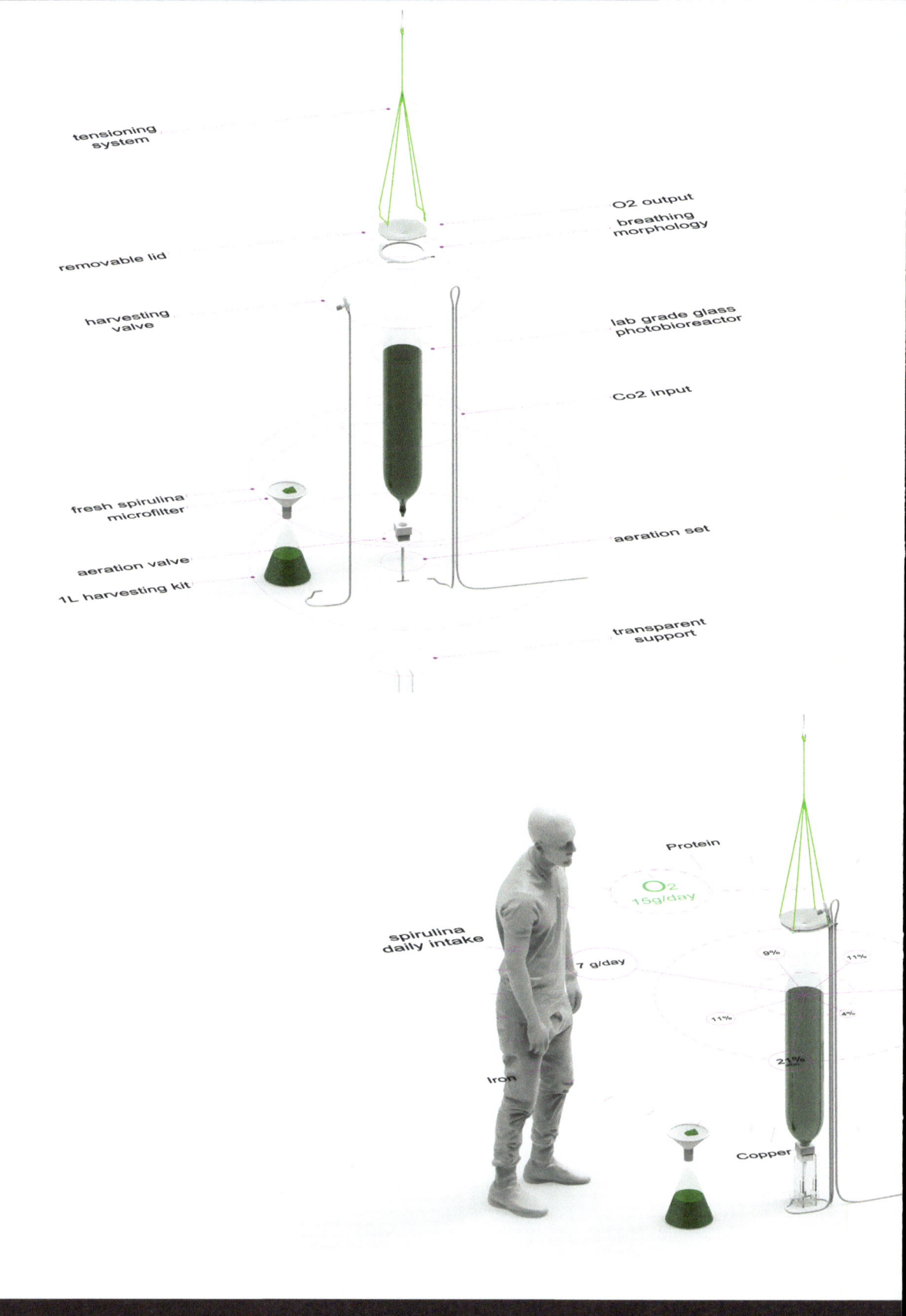

BioBombola diagrams explaining the kit of parts and its production cycles in terms of nutrients and air purification.

BioBombola can be easily assembled and dismantled, with zero waste. The photo-bioreactor adapts to any environment and any ceiling height, and it should be installed in a spot with diffuse sunlight or next to a grow lamp.

The product explores a visual and tactile way to introduce high-tech cultivation into the urban context by encouraging direct interaction. The authors' ambition is to redesign some of the logics that led us to the current health crisis. If we all engage in transforming air pollutants into highly nutritious aliments there will be fewer opportunities for viruses to exploit unsustainable food supply chains and for polluted atmospheres to cause us harm.

How will we live together?

1.01.02

Cultural and community realms 01 BIT.BIO.BOT

Claudia Pasquero and Marco Poletto, with their research partners, the Synthetic Landscape Lab at Innsbruck University and the Urban Morphogenesis Lab at the Bartlett UCL, were invited to participate in 'How will we live together?' the 17th International Architecture Exhibition, La Biennale di Venezia, curated by Hashim Sarkis, held in Venice from 22 May to 11 November 2021.

Within the 'As New Households' section of the Corderie dell'Arsenale, the studio built BIT.BIO.BOT, a 1:1 scale installation, an immersive experiment in the domestic cultivation of the urban microbiome. The experimental space is designed to test the coexistence between human and non-human organisms in the post-pandemic urban sphere. The installation is an urban laboratory that combines advanced architecture with microbiology to build an artificial habitat, managed by a collection of systems, which enable the collective cultivation of microalgae in the domestic realm.

The core biological mechanisms of BIT.BIO.BOT is the process of photosynthesis powered by the sun and by the metabolism of living cultures of Spirulina platensis and Chlorella spp. These living organisms are among the oldest on Earth and have developed a unique biological intelligence that enables them to convert solar radiation into actual oxygen and biomass with unrivalled efficiency. This advanced architectural system is the result of 10 years of bio-digital design research, developed by ecoLogicStudio and combining computational design strategies (BIT) with proprietary fabrication techniques (BOT) to implement a collective microbiological cultivation protocol (BIO).

Cover page: Envisioning the BIT.BIO.BOT algae farming village in the Venice lagoon. Venice Architecture Biennale 2021

This page: Intensive domestic cultivation of Chlorella spp. in a Vertical Algae Garden, at the Corderie dell'Arsenale. Venice Architecture Biennale 2021

Close-up view of the Living Cladding and the Vertical Garden creating new spatial typologies for domestic spaces within the Corderie dell'Arsenale. Venice Architecture Biennale 2021

BIT.BIO.BOT is composed of multiple interconnected systems: the 'Living Cladding', the 'Vertical Garden' and the 'Convivium'.

Living Cladding redefines the limits between human and non-human realms and between architectural indoor and outdoor. It is composed of ten PhotoSynthetica™ curtains. The unique version unveiled at the Corderie features a morphological pattern inspired by the surrounding brick walls, highlighting the microbiological nature of the Venetian architectural fabric. Furthermore, its articulation increases the interaction between microalgae growth in the bio-gel medium, and the environment, as well as their screening and shading potential. Each curtain is three metres high and one metre wide, and features 35 metres of digital welding, which forms a cavity capable of containing 7 litres of microalgae cultures.

The Vertical Garden creates a thick buffer zone located between the Living Cladding and the Convivium, dedicated to an intensive model of vertical algae farming. On a three-metre tall and completely reversible stainless steel structure, it hosts 15 BioBombola, the proprietary domestic algae garden system developed by ecoLogicStudio in response to the challenges of the first COVID-19 lockdown in spring 2020. Each unit in the Vertical Garden is made of laboratory-grade borosilicate glass and 3D-printed bioplastic components, and hosts 10 litres of microalgae cultures in a highly efficient growing medium. Algae can be independently harvested from each unit several times per week to collect up to one hundred grams of biomass, which is the daily recommended protein intake of a family of four. While active in production, the Vertical Garden is able to absorb CO_2 at a rate equal to that of three large mature trees, providing a clear path to carbon neutrality in architecture.

Vertical Garden assembly detail with bubbling Chlorella culture. The image highlights the complete reversibility of each unit which can be easily dismantled and reassembled in new locations and configurations.

The Convivium as a new culinary landscape at the Corderie dell'Arsenale, Venice Architecture Biennale 2021.

The Convivium is a space for sharing and for collective experimentation on the future of food. It takes the form of a 2x2 metre table, which hosts 36 unique pieces of crystal glassware designed to encourage experimentation in the consumption of the freshly harvested Chlorella and Spirulina cells, both of which are among the most nutritious organisms on Earth.

The crystal glassware is designed by ecoLogicStudio and 3D printed by Swarovski with its ground-breaking glass printing technology. Each piece is made of delicately fused glass layers, arranged in a matrix that algorithmically follows the morphogenesis of microalgae cells, thus generating a variety of visual patterns. These offer a multiplicity of opportunities to observe, transform and taste microalgae as part of a new culinary landscape.

The entire installation, as well as the individual systems, are fully reversible. At the end of the Architecture Biennale, in November 2021, each unit has become part of an educational algae garden project in collaboration with the ZKM Media Art Museum in Karlsruhe.

> "Each phase of the project including its conception, its fabrication, cultivation and post-Biennale re-functionalization, contributes to this overall experiment in coexistence—among us humans, and within an extended milieu of non-human systems."
>
> — Dr. Marco Poletto, co-founder of ecoLogicStudio

Detail of the 3D-printed glassware in the Convivium highlighting a catalogue of morphological variations.

Experiencing the future of food. An algae drink is being prepared in the 3D-printed glassware engaging the visitors of the Venice Architecture Biennale in a new culinary experience.

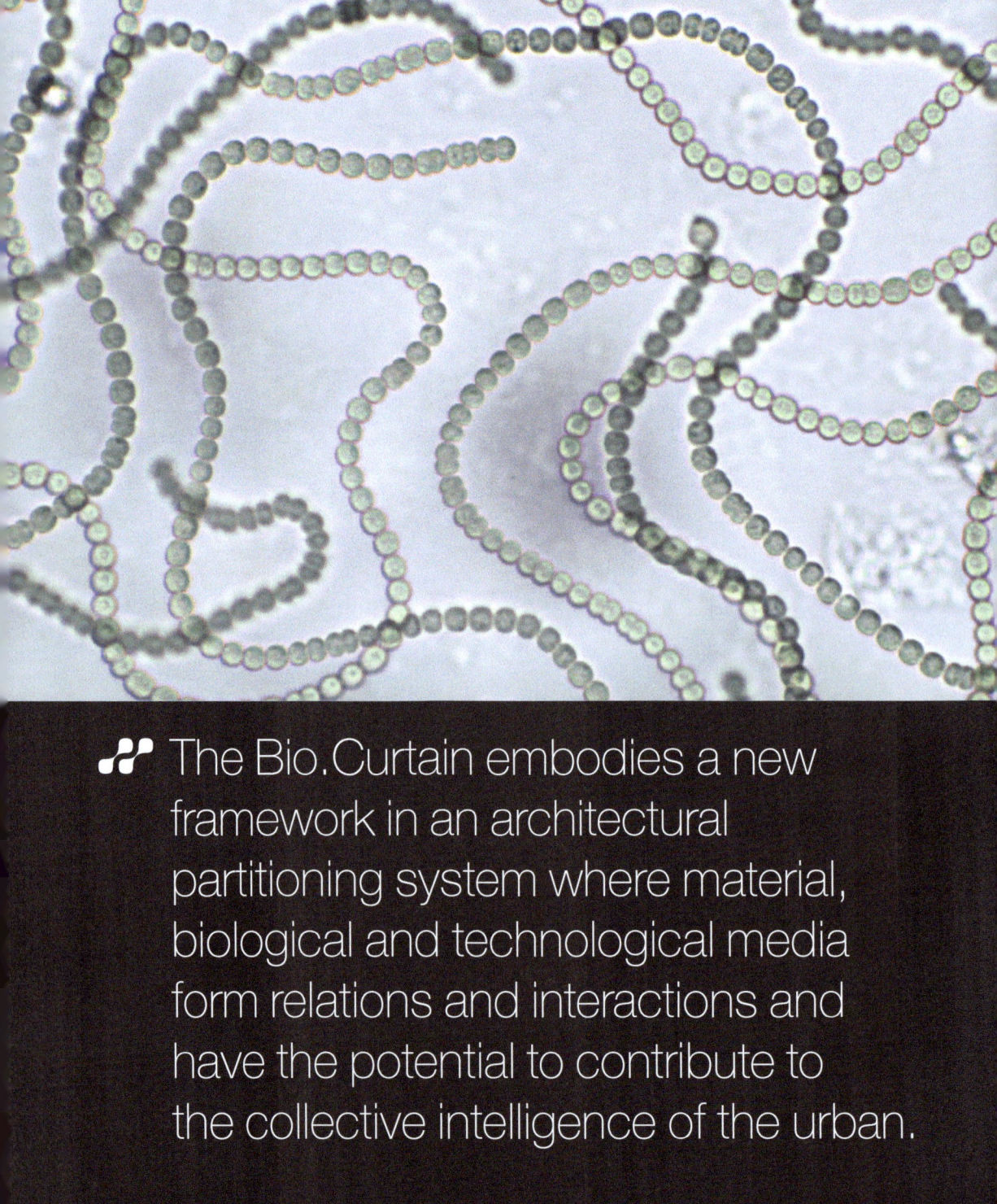

The Bio.Curtain embodies a new framework in an architectural partitioning system where material, biological and technological media form relations and interactions and have the potential to contribute to the collective intelligence of the urban.

1.01.03

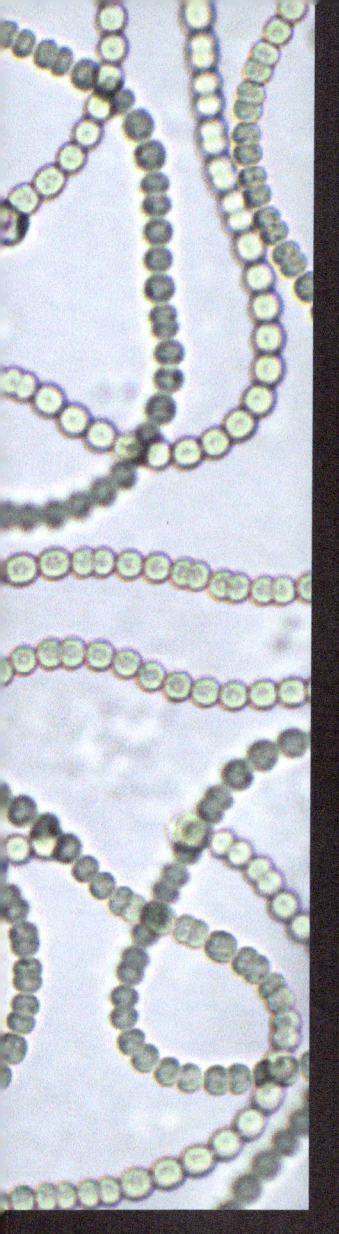

Cultural and community realms 02 Storytelling Bio.Curtain

Terézia Grešková

Architecture has the potential to become part of our extended cognitive system, a shift that affects us on a personal level as it touches our sense of responsibility in new and unknown ways. This becomes apparent while looking at the complexity of the current ecological crisis through the design lens of photosynthetic architecture, a perspective that directly triggers a larger discourse on the meaning of human perception, individual and collective intelligence.

Developed within the framework of doctoral research by Terézia Grešková at the Synthetic Landscape Lab, and as a result of a long-standing collaboration with ecoLogicStudio, the project 'Storytelling Bio.Curtain' proposes a unique architectural interface to articulate this shift of perspective. The project directly tests the functionality of technologies of photosynthetic architectures and specifically the photosynthetic algal bio-curtains, examining the behaviour of various strains of algae growing in bio-gel medium. In particular the project reflects upon our deep cognitive response mechanisms, directly influencing our perception of space.

What emerges is the notion of extended performance. The algal ability to metabolise pollutants and carbon dioxide, and to release oxygen back into the atmosphere, is directly experienced by the human organism through breathing. A conversation unfolds that can be directly explored through innovative digital sensing apparatus. The project creates a direct link that allows us to study the living organisms and their effect on the human self simultaneously. Crucially, recognising human experiences as a relevant part of our surrounding environment contributes to the development of an ecology of emotional freedom. Furthermore, evolving our individual human emotional state instils within us collective behavioural patterns of a shared biosphere.

This hypothesis was tested as part of an immersive installation 'Storytelling Bio.Curtain' via biometric mapping using galvanic skin response (GSR). In the installation, GSR sensors monitor emotional responses to the projected narrative of a Cyber-Yogi, a fictional character who does not neglect science and uses technology in order to deepen their knowledge about themselves and the world around them. At the same time, the Cyber-Yogi is a scientist who does not neglect their own emotions, but uses intuition to expand their own ability to perceive and generate inner knowledge. This enables them to extend their mind beyond the physical envelope. A Cyber-Yogi does not think of the skin as the outer layer separating themself from their surroundings.

While investigating these fundamental questions about the nature of human perception, the project starts from the assumption that the mind does not have an a priori location or place of origin and that we may, as L. Malafouris would say:

> "assume that the stuff of mind do not exist only inside the head but can be found also, if not primarily, inside the world" (Malafouris, 2019).

When we experience an arousing stimulus, our body generates a variety of psychophysical responses, that are usually in correlation with environmental pressures, such as traffic load, startling noise, environmental pollution and so on. Even a disturbing thought triggered by memory affects our internal emotional wellbeing and can be picked up by the sensors as a data pattern. Therefore, as we engage with the world, our bodies react with unconscious signals and we can measure the vulnerability of our experience. We can reveal the potential to enhance our own contribution to the world surrounding us and develop a sense of compassion.

Cover page: Microscopic photograph of living cultures of Nostoc taken by Synthetic Landscape Lab Innsbruck University.

This page: Architecture of Experience. The Storytelling Bio.Curtain exhibited at the Potenziale3 in Innsbruck, 2021.

The AirBubble exhibition at the Copernicus Science Centre in Warsaw, Poland.

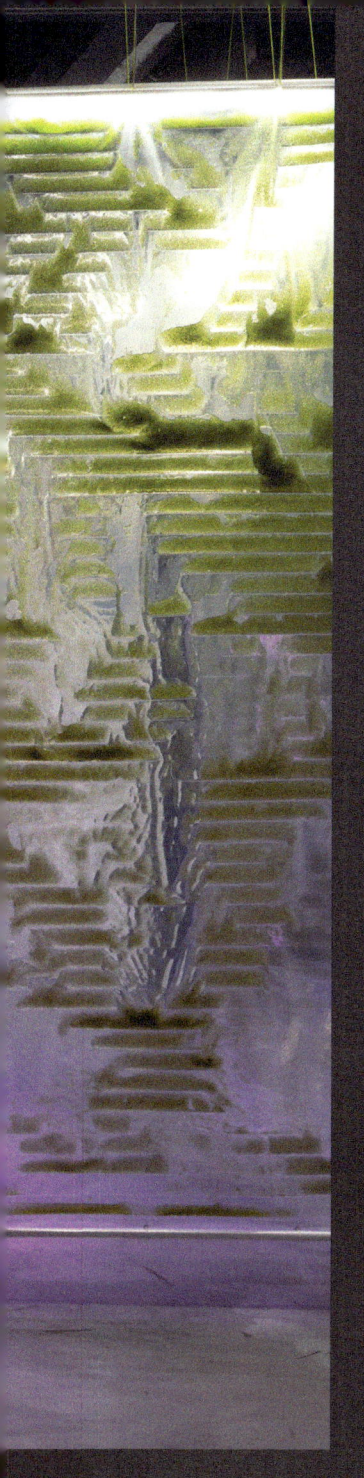

From the Latin *compati*, compassion means literally 'to suffer with'. However, compassion here is not only about feeling the pain of the other, but especially it is a motivation to relieve that pain by taking an active part in its circular metabolism. This moves us beyond a perception of the world as dominated by relationships of causes and effects, beyond searching for the 'guilty' and those 'responsible' for the current global instability. It dissolves this anthropocentric perspective and instils in us the ability to answer questions about the systemic nature of the 'Urbansphere', the global apparatus of contemporary urbanity.

As Lisa Feldman Barrett might say, emotions only seem to happen to us, but in fact are made by us (Barrett, 2017). This does not mean that emotions are an illusion or that bodily responses are random. It means that on different occasions, in different contexts, in different studies, within the same individual and across different individuals, the same emotion category involves different bodily responses. Therefore, the Storytelling Bio.Curtain project proposes the Cyber-Yogi practice as an instrument of self-inquiry, concluding that variation, not uniformity, is the norm. As such, architectural experimentation requires a conscious shift from designing carefully crafted artefacts set in a certain space towards a distributed multitude of dynamic processes unfolding in time (Pasquero, Poletto, Greskova, 2020).

This brings us to a further revelation of what we may call the utopian struggle of science, to see the present world as a transparent, entirely visible and entirely grasped object. Such an attempt could be defined as almost mystical, pursued with a great sense of positivism that erases the blank spaces of uncertainties. For, as Marco Poletto points out, smartness without participation leads to fragile and control-obsessive regimes which in turn favour the emergence of ideological approaches such as eco-conservativism. That is why we favour an engagement with the urban landscape as a real test bed for a participatory framework of bio-digital design, rather than simply a quest for the application of new green-tech devices (Poletto, 2018).

The Storytelling Bio.Curtain embodies this new framework in an architectural partitioning system where material, biological and technological media form relations and interactions and have the potential to contribute to the collective intelligence of the urban.

♪♪ A prototype for a future dwelling, energetic self-sufficient, carbon neutral and nutritionally circular. A delightful space to test a new kind of symbiosis between man and cyanobacteria.

1.01.04

Retail and showroom realms
Bio.Tech Hut

The Bio.Tech Hut was conceived and designed by ecoLogicStudio and built with Adunic in Astana, Kazakhstan for EXPO2017. It is currently part of the museum of future energy.

The concept of Bio.Tech Hut explores the anthropological relationship between man and the natural environment in the Anthropocene age. Such relationship was first explored in the concept of the 'primitive hut' originally brought to life in the mid-1700s, and further developed by Marc-Antoine Laugier in his work *An Essay on Architecture*. It contends that the ideal architectural form embodies what is natural and intrinsic. This was one of the first significant attempts to theorise architectural knowledge both scientifically and philosophically. Today as biotech is redefining the boundaries of the natural we ask ourselves what repercussions it will have on our future life, the architecture of domestic spaces and the use of natural resources.

The Bio.Tech Hut is arranged into three fluidly interconnected environments, loosely embodying the fundamental programmes of a dwelling. The first space is the 'Lab', where new species of microorganisms are domesticated and engineered into artificial environments of cultivation, patterns of growth and material assemblies. It is a space of science and rationality.

From here two interwoven corridors lead into the 'Living Hut', the core space of the dwelling. It is a space of artistic experience divided into two rooms. One is flooded with natural light and hosts H.O.R.T.U.S., Hydro Organisms Responsive

It is inhabited by photosynthetic colonies of cyanobacteria, which visitors are called on to nurture with their exhalations of carbon dioxide; in turn they receive oxygen and the growing biomass. A form of domestic symbiosis emerges, one that seeks augmented interaction between the world of micro-particles and the space of human inhabitation. H.O.R.T.U.S. and its convolutedness are the spatial embodiment of this relationship between energy and form.

Next is the bio-light room, a space of calm, slowness and darkness; the only visible light is emitted by bioluminescent bacteria when they are shaken, and the room is oxygenated by the dwelling's air circulation system.

The Living Hut leads into the third and more open environment, the 'Garden Hut' where production is concentrated. The algae photo-bioreactor room is completely clad in growing microalgae. These highly photosynthetic microorganisms generate biomass and oxygen while absorbing carbon dioxide. Central to this part of the dwelling is the harvest area, a communal space of craft, for the processing and transformation of biomass in food and electricity.

The structure and the cladding of the Bio.Tech Hut integrate multiple levels of performance. The programmatic articulation of the dwelling and its performance are deeply interrelated, expressing architecturally the notion of symbiosis between human and non-human dwellers.

The photo-bioreactor cladding, designed by ecoLogicStudio in collaboration with LGem, is developed from a revolutionary system that uses a high velocity air flow to lift the living medium up into laboratory-grade glass tubes. The fast air stream creates eddies on the liquid surface, and a unique stirring effect that gives the desired O2–CO2 exchange. Also, the bubbles keep the wall of the tubes clear improving the quality and productivity of the culture. The fluid then descends by gravity to complete the loop. Multiple loops are coiled around the garden and the living spaces, thus becoming defining architectural elements. This was achieved by individually designing and numbering each pipe in the system. This kind of integrated system is a world first.

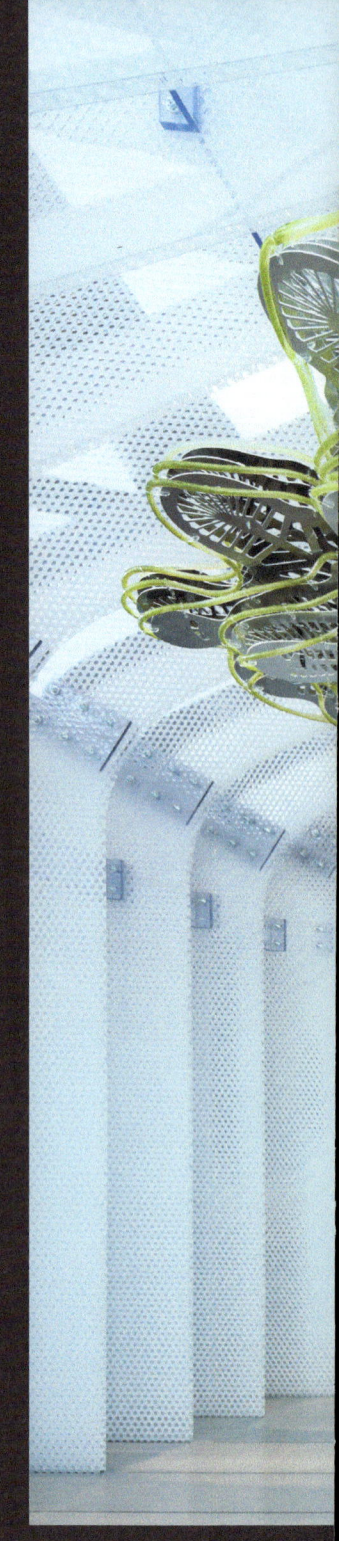

Cover page: The living space of the Bio.Tech Hut at the museum of future energy at Astana, visitors experiencing a new kind of symbiosis with algal microorganisms.

This page: H.O.R.T.U.S. hangs from the ceiling of the Living Hut as a cloud-like chandelier; its convolutedness responds to incoming light radiation. Algae circulate in loops around the structure and are manually oxygenated by visitors with pumps.

Detail of the laboratory-grade glass tubes containing different algae cultures and the honeycombed polycarbonate structure.

Structurally the coils are supported by a series of sectional frames in high performance honeycombed polycarbonate. The resulting structure is lightweight, fully recyclable and most importantly has the unique effect of scattering and enhancing the penetration of solar radiation deep into the Hut. This further enhances the growth of the microalgal organisms.

The Bio.Tech Hut is 180 metres square in plan and supports 1600 litres of living cultures of cyanobacteria in its lab-grade glass photo-bioreactors. In optimal conditions it is a test bed of self-sufficiency, demonstrating the feasibility of integrating such horticultural systems into high-end corporate retail and commercial environments. It proposes a new aesthetic dimension for carbon capturing and sustainable production that is profoundly urban and that will become part of the future fabric of our cities.

Ultimately, to achieve carbon neutrality it will not be sufficient to absorb carbon from the atmosphere. It is critical that cities and buildings enable the emergence of new circular economies so that food, products and energy can be delivered with little or no associated carbon footprint. Within PhotoSynthetica™, the living algae colonies do not simply absorb CO2 but they convert it into biomass, which is sugars, proteins and other nutrients, as well as oxygen.

These provide useful raw material for numerous production processes. By growing algae within the urban realm, PhotoSynthetica™ provides a platform for such production processes to take place on site and with zero harmful emissions.

View of the Garden Hut at the Museum of Future Energy, Astana. The Garden Hut is a communal and open space for the collective processing and transformation of biomass into food and biofuel.

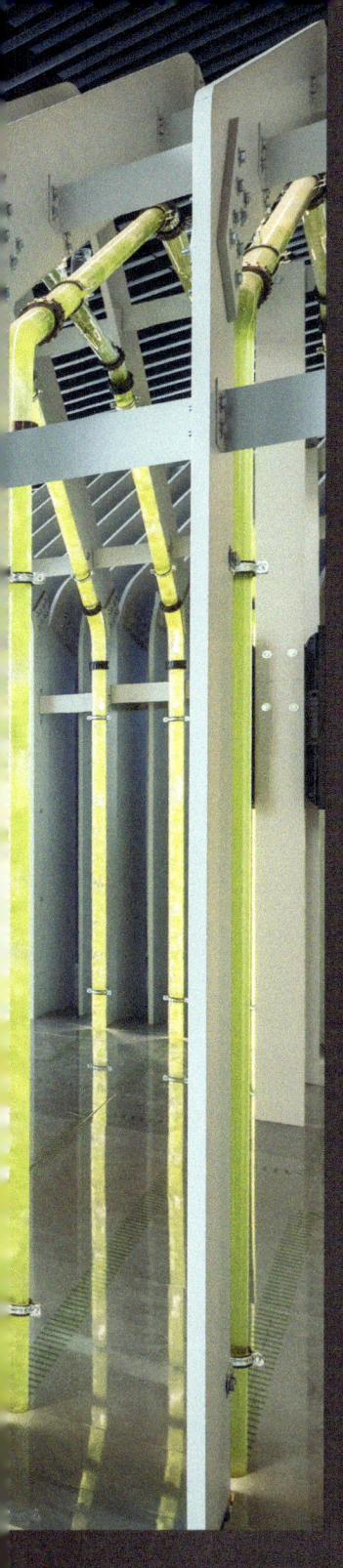

Oxygenation

The average adult inhales 550 litres of O2 per day and it takes 8 large trees to produce that amount daily. The average adult also exhales 1kg of CO2 per day and it takes approximately 16 large trees to absorb that. The Bio.Tech Hut hosts a total of 1,600 litres of living cultures. In optimal conditions these absorb 2kg of CO2 per day and produce 1.5kg of O2. Bio.Tech Hut can absorb an amount of CO2 equivalent to that of 32 large trees.

Bio-fuels and energetic self-sufficiency

On average the Bio.Tech Hut produces 1.12kg of dry algae per day. Green microalgae like Schiochytrium or diatoms like *Phaeodactylum tricornutum* can contain up to 60% oil in dry weight. Hence Bio.Tech Hut can produce 672g of oil per day from which approximately 1kg of biofuel can be produced, releasing up to 37 megajoules (MJ) of energy or 10.3 KWh, which is enough to power a typical UK home.

Embodied energy and new urban metabolism

A microalgae like Chlorella contains up to 60% proteins, so every day the Bio.Tech Hut could produce 672g of proteins per day, enough to supply the recommended daily intake of 12 adults. An average cow provides 52kg of meat-based proteins and it takes 660 days to breed it. Therefore the Bio.Tech Hut produces the equivalent meat-based proteins of 8 cows. The breeding process consumes on average 2.682 MJ per day to do that, equivalent to the energy consumption of 70 average UK households.

Carbon neutrality

Every cow also produces methane equivalent to 4 tons of CO2 per year; for 8 cows that is 87kg per day. If we would switch to an algae protein diet the net contribution to carbon sequestration of the Bio.Tech Hut would be approx. 90kg per day. It takes 500 square metres of forest to do that.

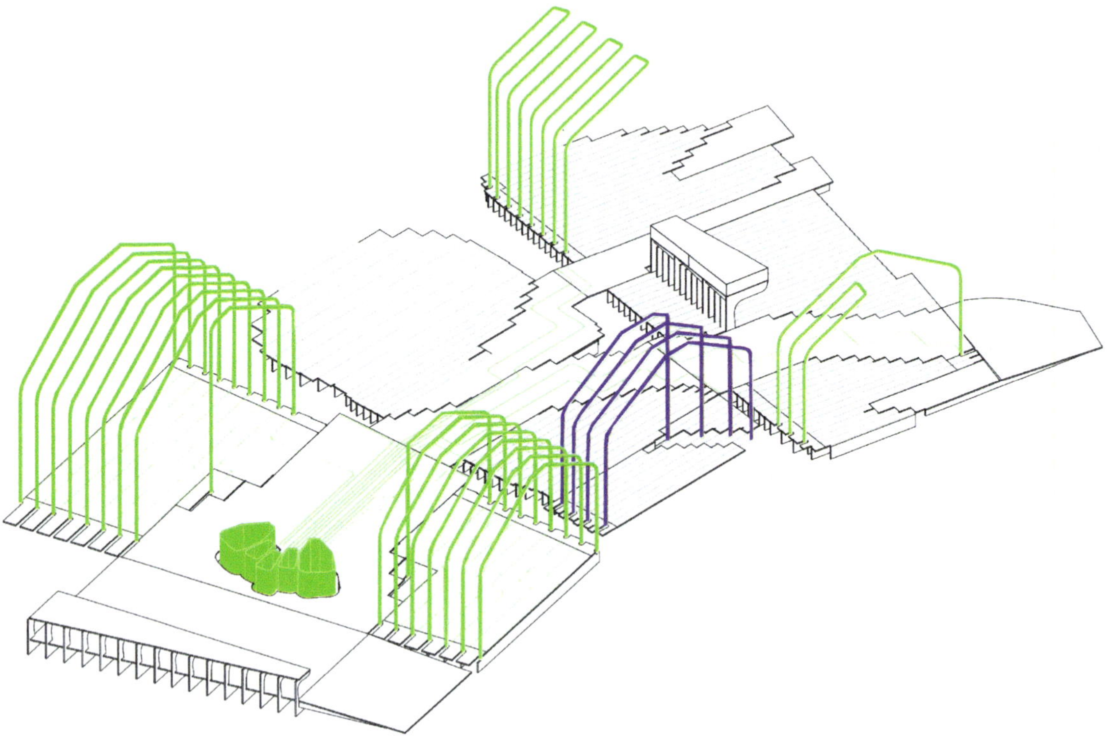

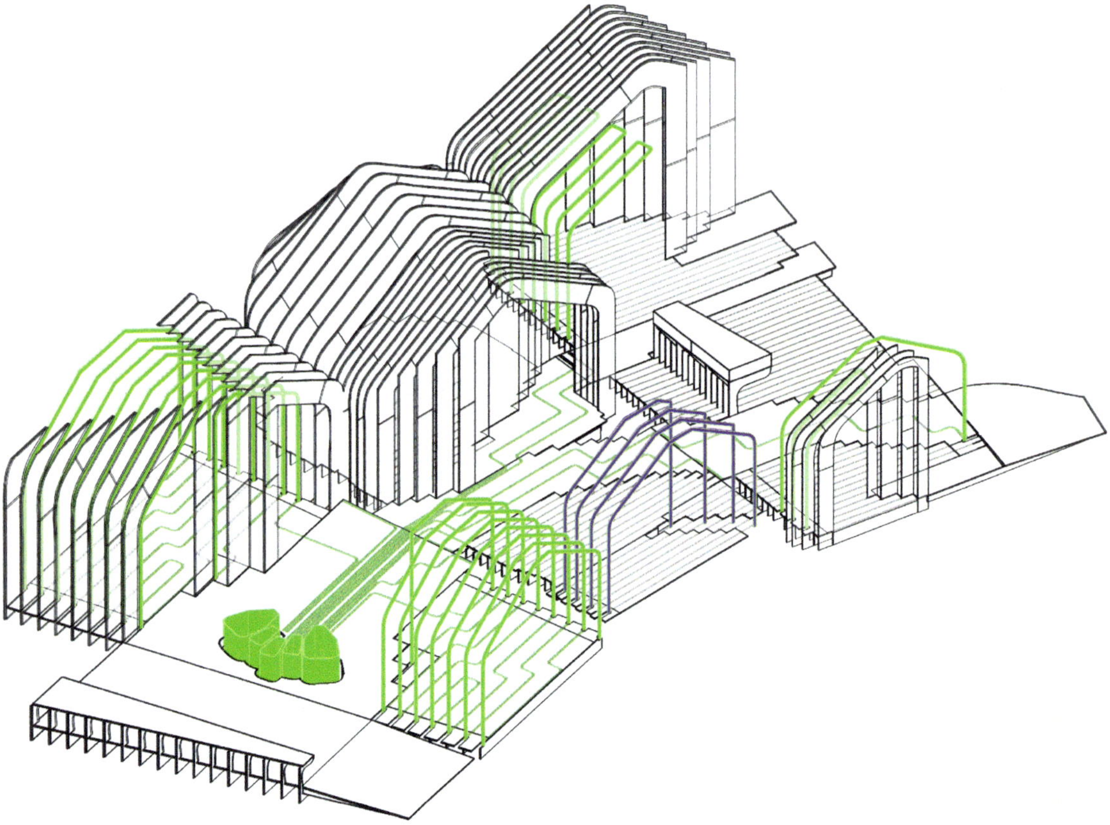

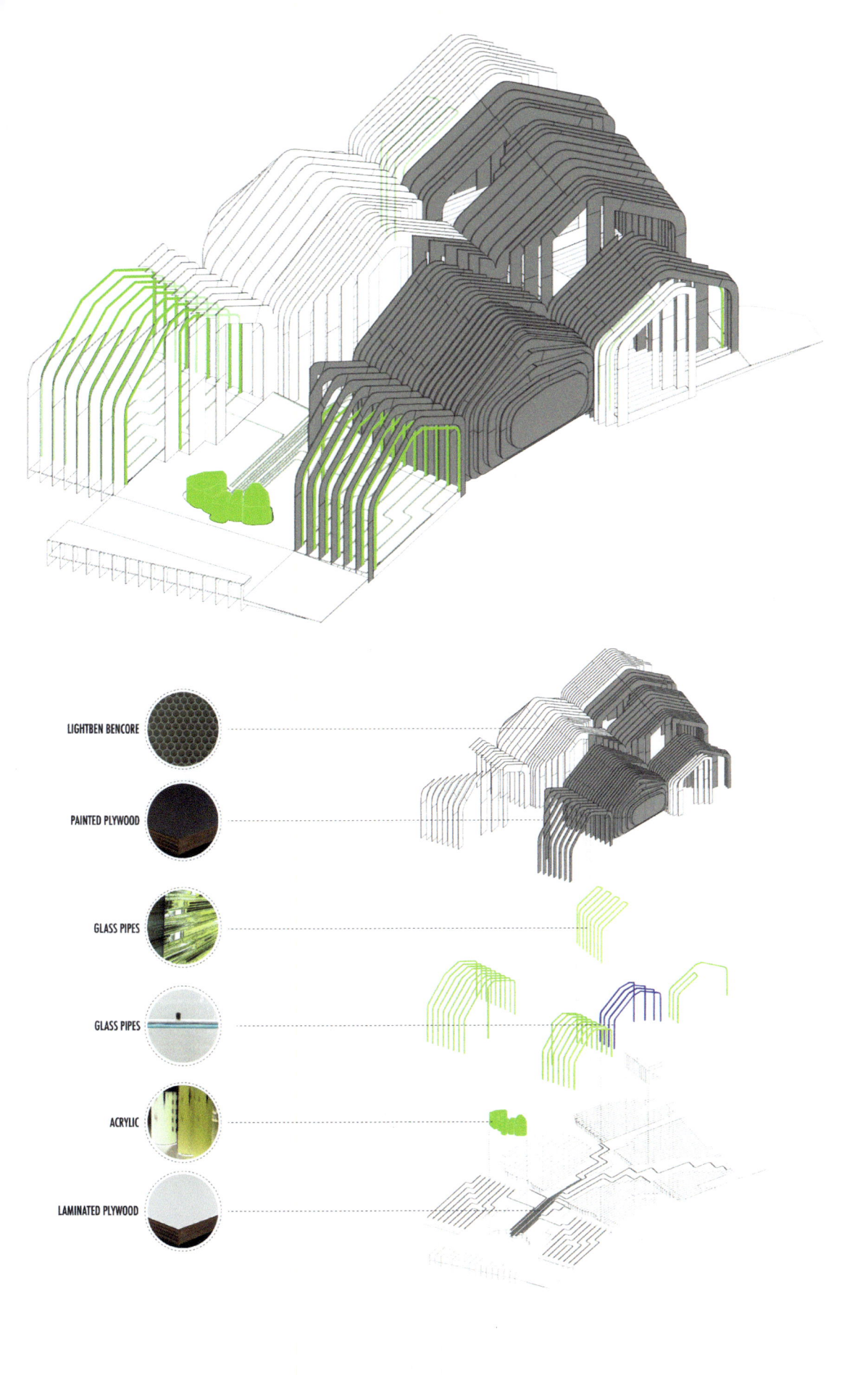

- LIGHTBEN BENCORE
- PAINTED PLYWOOD
- GLASS PIPES
- GLASS PIPES
- ACRYLIC
- LAMINATED PLYWOOD

Previous spread: Axonometric construction sequence of the Bio.Tech Hut highlighting the integration of architectural structural and microbiological systems.

This page: Detail of H.O.R.T.U.S. photosynthetic surface.

Front view of the entrance to the Garden Hut during WORLD EXPO 2017 in Astana, Kazakhstan.

♪♪ PhotoSynthetica Curtains use the power of algae to absorb carbon dioxide from the air, demonstrating how biotechnology can become integrated in our cities to help achieve carbon neutrality.

1.01.05

Architectural retrofitting 01
PhotoSynthetica Dublin

PhotoSynthetica Dublin, a large-scale urban curtain installation, was designed for Climate-KIC, the EU's most prominent climate innovation initiative, which aims to accelerate solutions to global climate change.

Conceived as an 'urban curtain', PhotoSyntheticaTM, was presented in Dublin during the week of Climate Innovation Summit 2018. It captures CO2 from the atmosphere and stores it in real time at a rate of approximately one kilo of CO2 per day, equivalent to that of 20 large trees. The innovative system could be integrated into existing and newly designed buildings.

Composed of sixteen 2 x 7 metre modules, the unique curtain prototype envelopes the first and second floor of the main façade of the Printworks building at Dublin Castle. Each module functions as a photo-bioreactor, a digitally designed and custom-made bioplastic container that utilises daylight to feed the living microalgal cultures and releases luminescent shades at night.

Cover page: Front view of the PhotoSynthetica, Dublin, wrapping the main facade of the Printworks building at Dublin Castle.

This spread: Diagram of the carbon capturing efficiency of the PhotoSynthetica Curtain compared to that of a mature tree, superimposed onto the main façade of the Printworks building.

Diagram of the photosynthetic functioning principle of the PhotoSynthetica Curtain superimposed onto the main facade of the Printworks building.

Unfiltered urban air is introduced at the bottom of the Photo-Synthetica façade and, while air bubbles naturally rise through the watery medium within the bioplastic photo-bioreactors, they come into contact with voracious microbes. CO2 molecules and air pollutants are captured and stored by the algae, and grow into biomass. This can be harvested and employed in the production of bioplastic raw material that constitutes the main building material of the photo-bioreactors. To culminate the process, freshly photosynthesised oxygen is released at the top of each façade unit of PhotoSynthetica™, and out into the urban microclimate.

Thanks to their serpentine design, the modules optimise the carbon sequestration process and the full curtain pattern is reminiscent of a large trading data chart that embodies Climate-KIC's commitment to promote new models to solve the global climate crisis.

Moreover, the PhotoSynthetica™ project seeks to symbolically embody a parallelism between the monetary carbon trading market and the transactions carried out by the molecules. The message is one of spatial convergence and connectivity between the financial marketplace of cyberspace and the relative organic molecular transactions in the biosphere.

Close-up view of the PhotoSynthetica Curtain as a retrofitting system of the 1980s Printworks building. The image demonstrates how biotechnology can become integrated in our cities to help achieve carbon neutrality.

Detail of the inoculation process of the PhotoSynthetica Curtain.

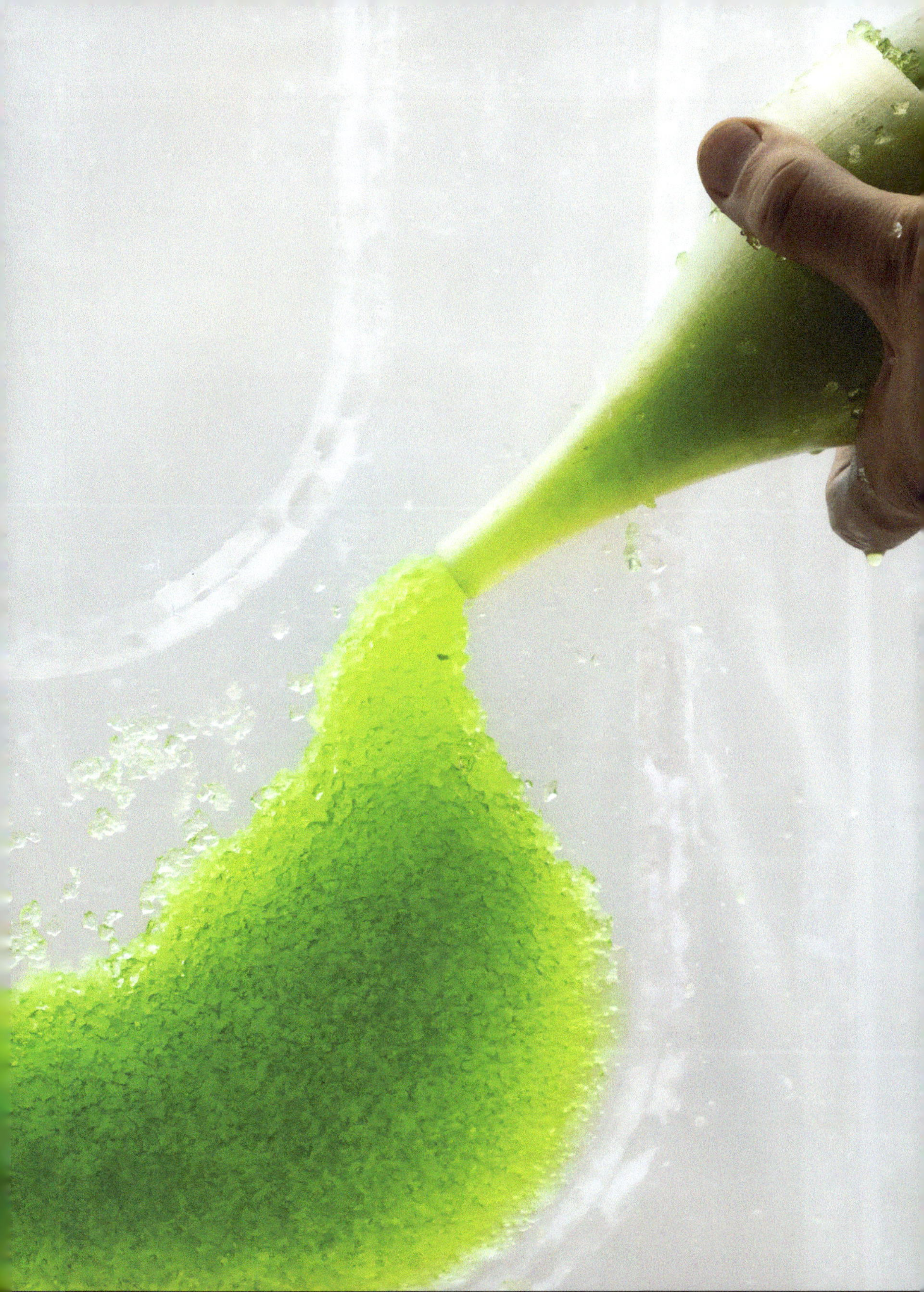

A soft and living building fabric that literally breathes.

1.01.06

Architectural retrofitting 02
PhotoSynthetica Helsinki

Historically one of the defining qualities of architecture is its permanence, its hardness and its resistance to change and movement. However, every building undergoes periods of more rapid change and evolution when its immovable body becomes softer and more malleable.

In this phase the body of architecture appears more akin to the human body or to the other soft and wet organisms that make up the living biosphere. What if we could take this temporary condition and transform it into an opportunity to rethink the fabric of architecture, wrapping it in a living membrane that would both protect and connect the body of architecture with the surrounding environment?

And what if this soft and wet skin could literally breathe, make use of the energy of the sun to capture molecules of carbon dioxide and release oxygen? And what if this carbon could be stored in a fibrous biomass to be used for future textiles?

This is the ambition behind the installation developed by ecoLogicStudio for the Nobility House in Helsinki, a prestigious historical building hosting Helsinki Fashion Week. The Nobility House Urban Curtain is made of 100 modules each containing 10 litres of photosynthetic microalgae. Six modules of the curtain capture the same amount of CO_2 as a mature tree transforming the House into a biotechnological micro-forest.

Cover page: Detail of the PhotoSynthetica Curtain components woven across the scaffolding structure erected on the main facade of the Nobility House Helsinki, in summer 2019.

This spread: The main facade of Nobility House in Helsinki is dressed with more than 100 modules of PhotoSynthetica Curtain woven together with yellow nylon ropes.

✸ This project tests the feasibility of building bio-factories where microalgae-based raw materials are grown efficiently on the building walls.

1.01.07

Architectural retrofitting 03
BioFactory

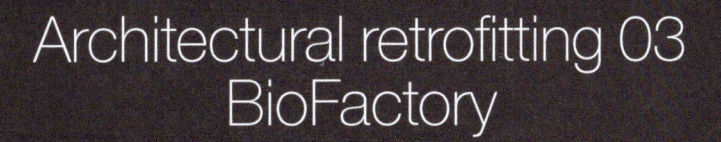

Through the application of the PhotoSynthetica™ technology this research project tests the feasibility of building bio-factories in the near future, where microalgae are grown efficiently on the factory walls.

The 'BioFactory' will thus implement locally circular economies of matter and energy. Microalgae will grow inside the photo-bioreactors while feeding on the CO_2 emissions of the factory. Freshly harvested biomass will then enter the factory supply chain to become a renewable and sustainable raw material for carbon neutral food products and packaging.

The system couples the unique biological intelligence of photosynthetic microalgae with the artificial intelligence of autonomous farming protocols. This circular process changes the rules of efficiency and makes bio-factories capable of self-regulation and learning. As the system grows, it becomes more resilient and ultimately evolves higher levels of productivity.

Algae exists in a large variety of species, each with unique properties that can co-evolve with the needs of different supply chains. They grow quickly and can adapt to extreme urban conditions while absorbing and re-metabolising most pollutants populating air and water. A BioFactory is characterised by a low-carbon construction process that will make the factory's cladding carbon negative over the span of its lifetime. Moreover, the system is mostly made of recyclable materials such as glass and ETFE. In particular, ETFE foil is an extremely durable and lightweight material which is 100% recyclable, and requires minimal energy for transportation and installation.

Cover page: Front view of the bio-factory photosynthetic wall module installed at Nestlé Portuguese headquarters in Lisbon

This page: The bio-factory wall module integrated in the kitchen and living area of the new Nestlé Portuguese headquarters in Lisbon. Employees can harvest fresh Spirulina daily.

The main entrance of the new Nestlé Portuguese headquarters in Lisbon featuring the integrated photosynthetic wall module

By integrating the units of photo-bioreactors in the building facades, the project creates an adjustable and intelligent shading system, thereby reducing buildings' cooling loads while allowing natural light in. At the same time adaptive or responsive shading also increases the psychological benefit for building users and employees. In fact, proximity to natural elements has been associated with a 15% improvement in wellbeing and creativity, and a 6% increase in productivity.

The BioFactory concept envisions a new space for production embedded in a systemic approach to true sustainability. The convergence of automation, artificial intelligence and biotechnology has the potential to transform the role of factory employees who gain unprecedented agency in the production process. New job definitions will reflect the interactions enabled by the system. From advanced computation to micro-farming, bio-design, future culinary activity and gardening, BioFactory fosters a proactive attitude exemplified by the fundamental aspects of cultivating, harvesting and processing living organic material.

Detail of the lab-grade glass photo-bioreactors and 3D-printed bioplastic caps of the bio-factory wall module.

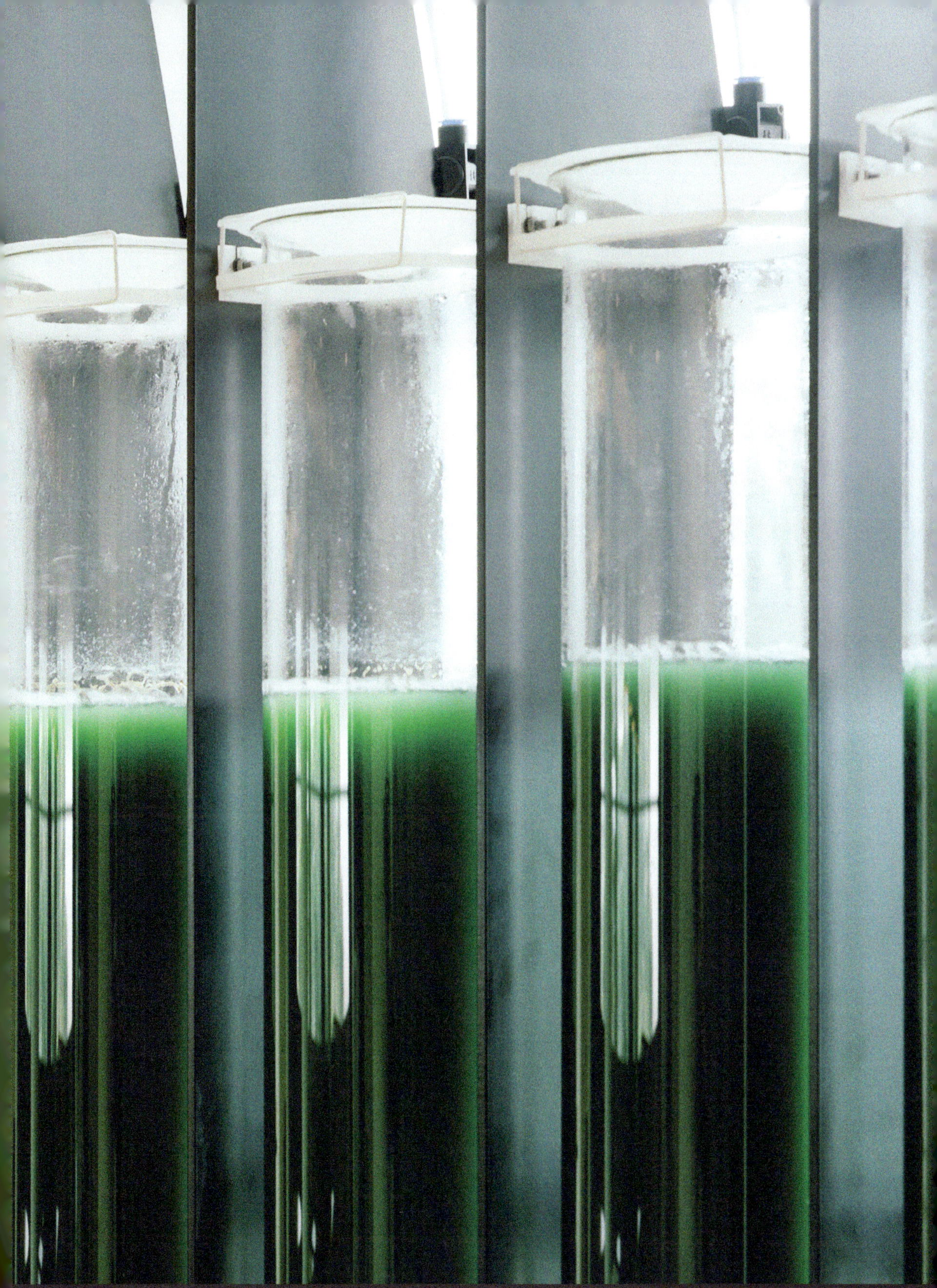

Interior view of a bio-factory where autonomous algae cultivation systems and human cyber-gardeners work together to create circular systems of production in the food industry.

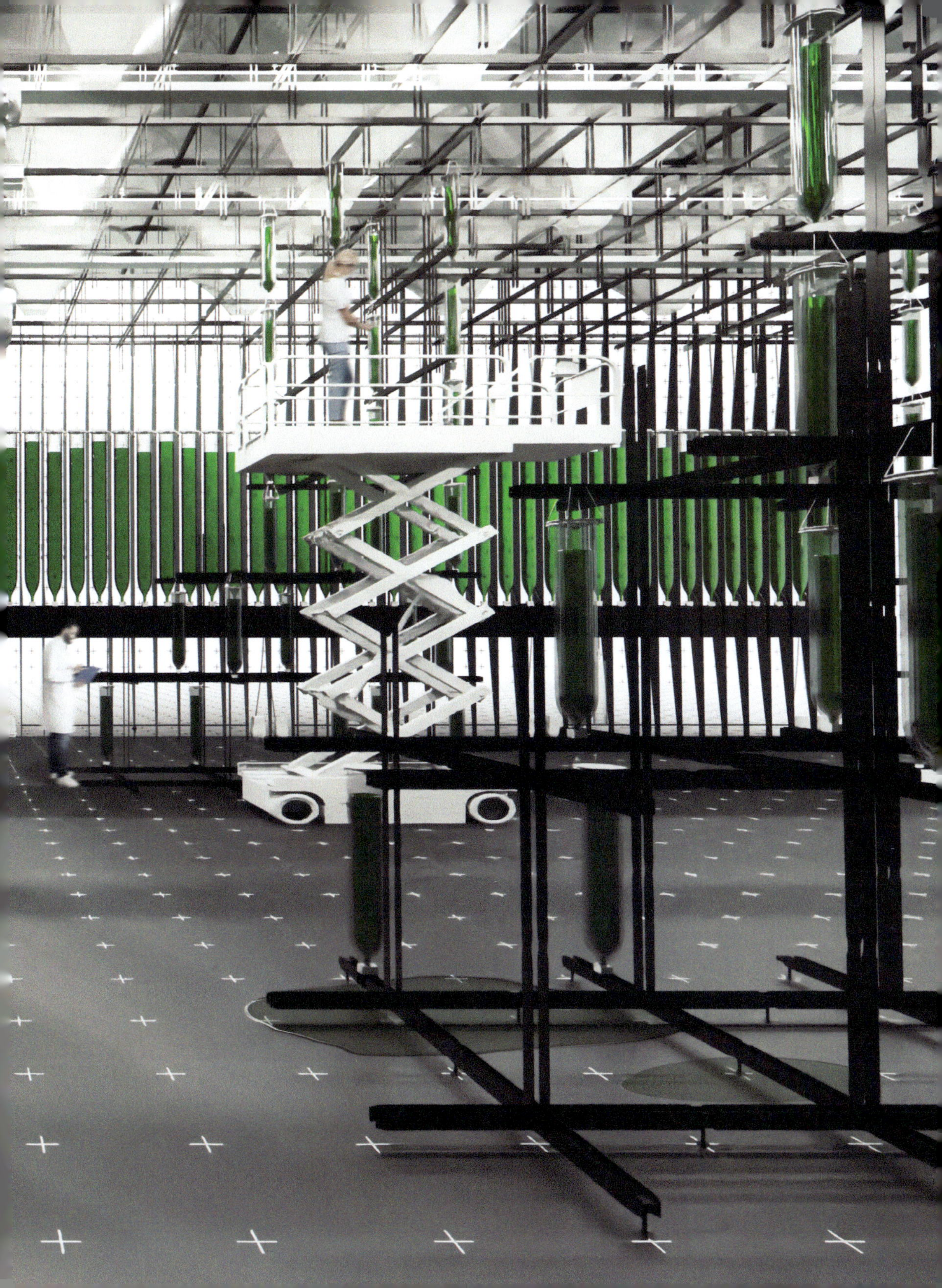

🎵 With AirBubble we bring to life the world's first biotechnological playground to integrate air-purifying microalgae.

1.01.08

The public realm
AirBubble

'AirBubble' creates a purified microclimate for children to play in, a true bubble of clean air in the centre of Warsaw (Poland). The project is located within the public green space outside the Copernicus Science Centre (Centrum Nauki Kopernik), a site which will also host a dedicated exhibition illustrating the design innovation behind the invention of AirBubble. The playground integrates the PhotoSynthetica™ technology for the advanced integration of photosynthesis in the built environment.

> "There is untapped value in bringing the bio-intelligence of natural systems into cities, turning buildings into living machines that produce energy, store CO2 and clean the air. To achieve this, we need to think about the living world as a part of the current digital revolution: nature becomes part of a new bio-smart infrastructure."
>
> — Dr. Marco Poletto, co-founder of ecoLogicStudio

According to the World Health Organization, air pollution is the biggest single global health threat. Warsaw was selected as the first activation for this project as it is one of the most polluted cities in Europe.

AirBubble invents a new architectural typology. It incorporates a cylindrical timber structure wrapped in an ETFE membrane protecting 52 glass algae reactors. This creates a real urban algae greenhouse. The space is equipped with ropes, foot pumps and bouncy spheres, and can function as both playground and outdoor classroom. The white bubbling noise of the algae gardening system masks the surrounding urban noise to provide a calming atmosphere in which to play and interact.

The filtering process is enhanced by the architectural morphology of the playground structure. The ETFE membrane, an evolution of the PhotoSynthetica urban curtain system presented in Dublin in 2018, controls the microclimate inside AirBubble. The inverted conical roof membrane further stimulates the air recirculation and natural ventilation, which in turn keeps the play area clean.

The AirBubble monitoring system integrates urban air pollution sensors and is connected to a data-processing platform capable of comparing measurements in real time and of highlighting the air quality index for six core pollutants: fine particulate PM2.5 and PM10, ground level ozone (O_3), nitrogen dioxide (NO_2), sulphur dioxide (SO_2) and carbon monoxide (CO). AirBubble is capable of absorbing 97% of the nitrogen and 75% of the particulate matter in the air. Data collected throughout summer 2021 shows that concentrations of PM2.5 within the playground have fallen well within the recommended WHO limits (green zone, AQI below 20).

Cover page: Children playing within the purified microclimate of the AirBubble in Warsaw

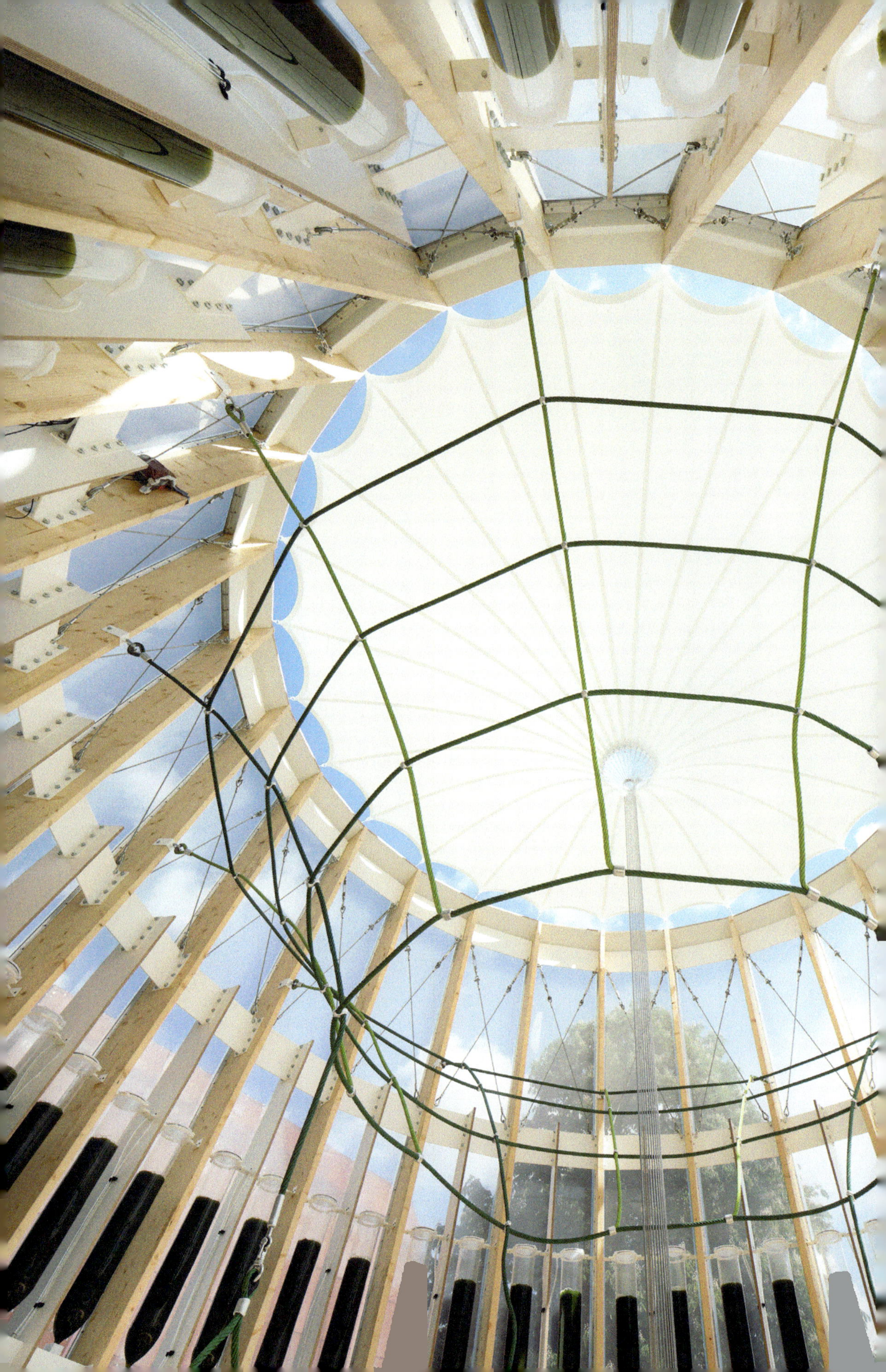

The peak reduction rate is an impressive 83%. This figure has been calculated by comparing readings from a pollution sensor located outside AirBubble with real-time data feeds from a monitoring device placed inside. The monitoring phase will continue in the future at several other locations as well as in other cities, to verify these promising achievements over a longer period of time, under different climatic conditions and patterns of use. Over the next few years, the AirBubble project will become a unique urban laboratory, a test bed of applied biotechnology to tackle air pollution and to mitigate its effects on children's health.

The AirBubble hosts 52 large bioreactors in borosilicate glass which contain 520 litres of living green Chlorella spp. algae cultures that can filter a flow of polluted air at 200 litres per minute. While the liquid medium washes particles, the algae actively eat the polluting molecules as well as carbon dioxide to then release fresh clean oxygen. The purifying process is powered by solar energy and children's playfulness. Children can interact by jumping on four water foot pumps positioned on the ground while balancing on the bouncy bubbles and the internal rope system.

"This playground needs two sources of power: solar energy and kids' instinctive drive to explore and to play. These constitute the inexhaustible and renewable fuels of the AirBubble that can be obtained effortlessly. The AirBubble is the trigger of a process that can only grow and multiply its beneficial effects towards future generations. It's all in our hands—we are responsible for our health and climate."

— Prof. Claudia Pasquero, co-founder of ecoLogicStudio

To mark the inauguration of the first ever AirBubble, ecoLogicStudio has designed an interactive multimedia exhibition at the Copernicus Science Centre (Centrum Nauki Kopernik) illustrating the architectural innovation and biotechnology at the heart of the playground system. The exhibition was composed of three areas: 'Monitor', exploring urban air pollution from the unique perspective of children's health; 'Purify the air', revealing the powerful symbiosis of architecture and living microorganisms—this section focused on the filtration and re-metabolisation of air pollutants; and 'Breathe', engaging children in an interactive demonstration of how to heal their bodies and our cities through the energy of play and breathing clean air.

Previous page: Fisheye view of interior of AirBubble. The playground appears as an outdoor room surrounded by the air-purifying Chlorella ring, composed of 520 litres of living cultures of microalgae. The inverted conical roof membrane further stimulates the air recirculation and natural ventilation and guarantees optimal microclimatic conditions.

This page: AirBubble at night in front of the Copernicus Science Centre in Warsaw in summer 2021.

AirBubble in front of the Copernicus science Centre in Warsaw.

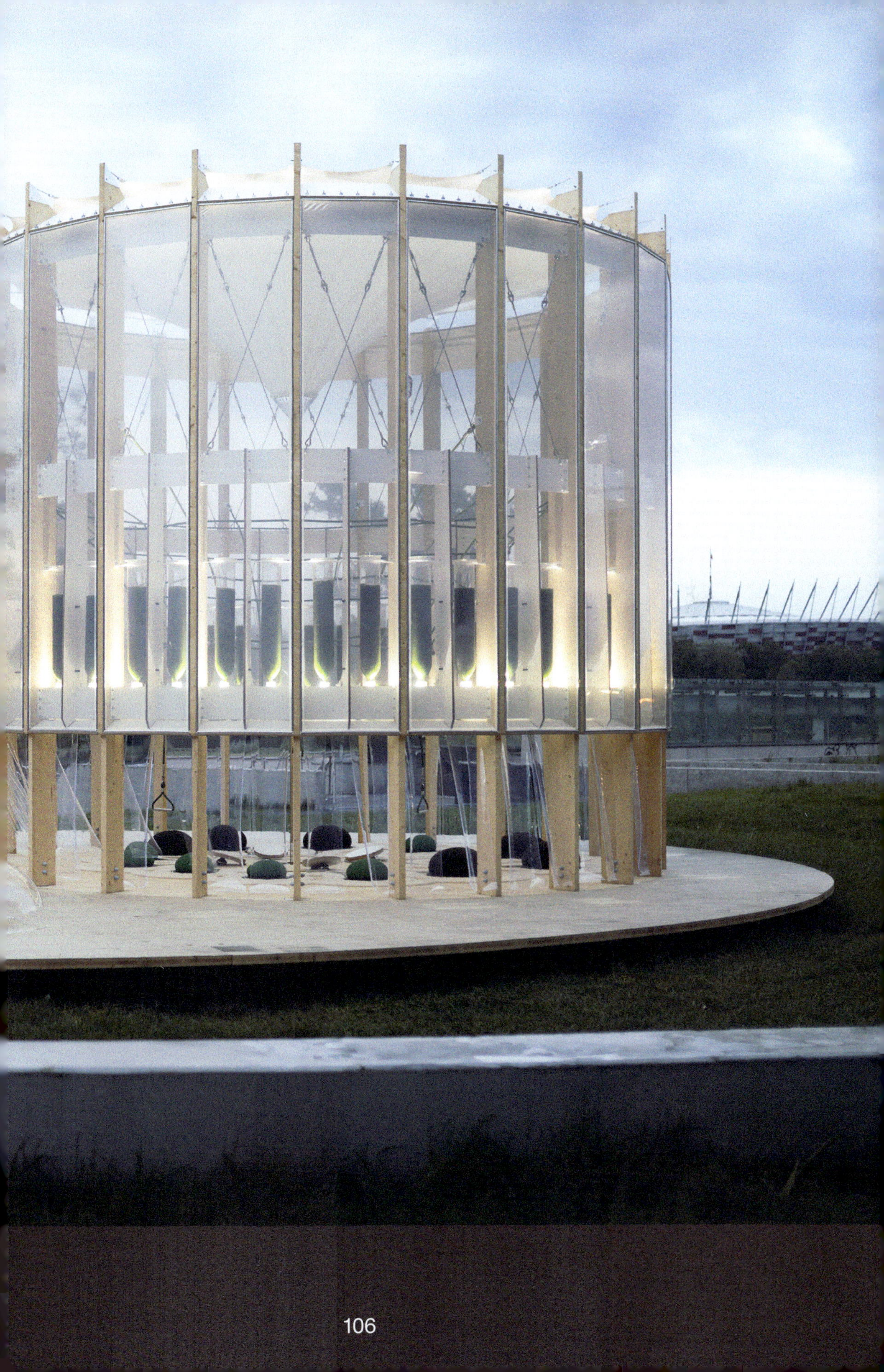

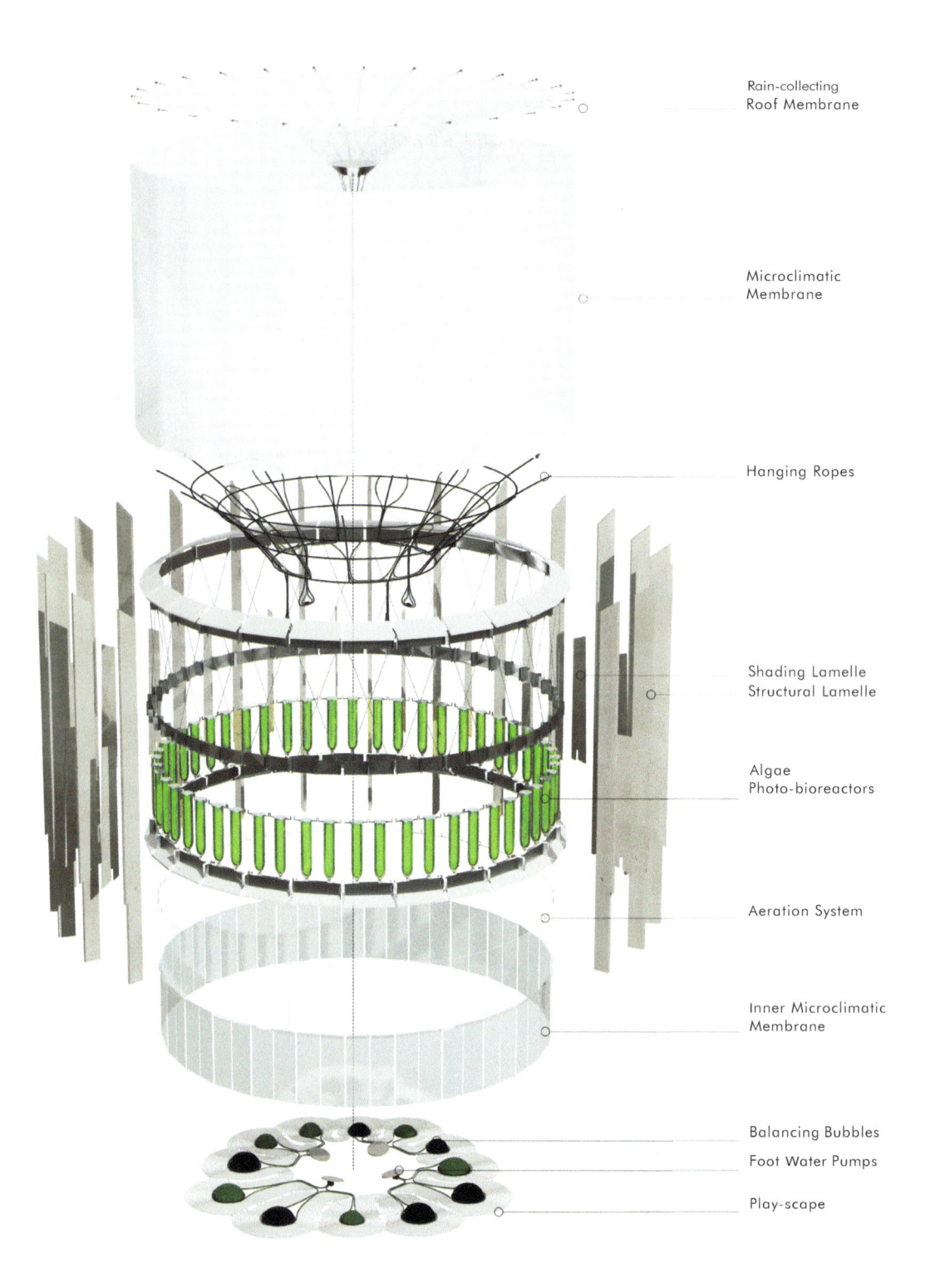

Exploded diagram of AirBubble highlighting its core architectural components and systemic integration of the biotechnological layers.

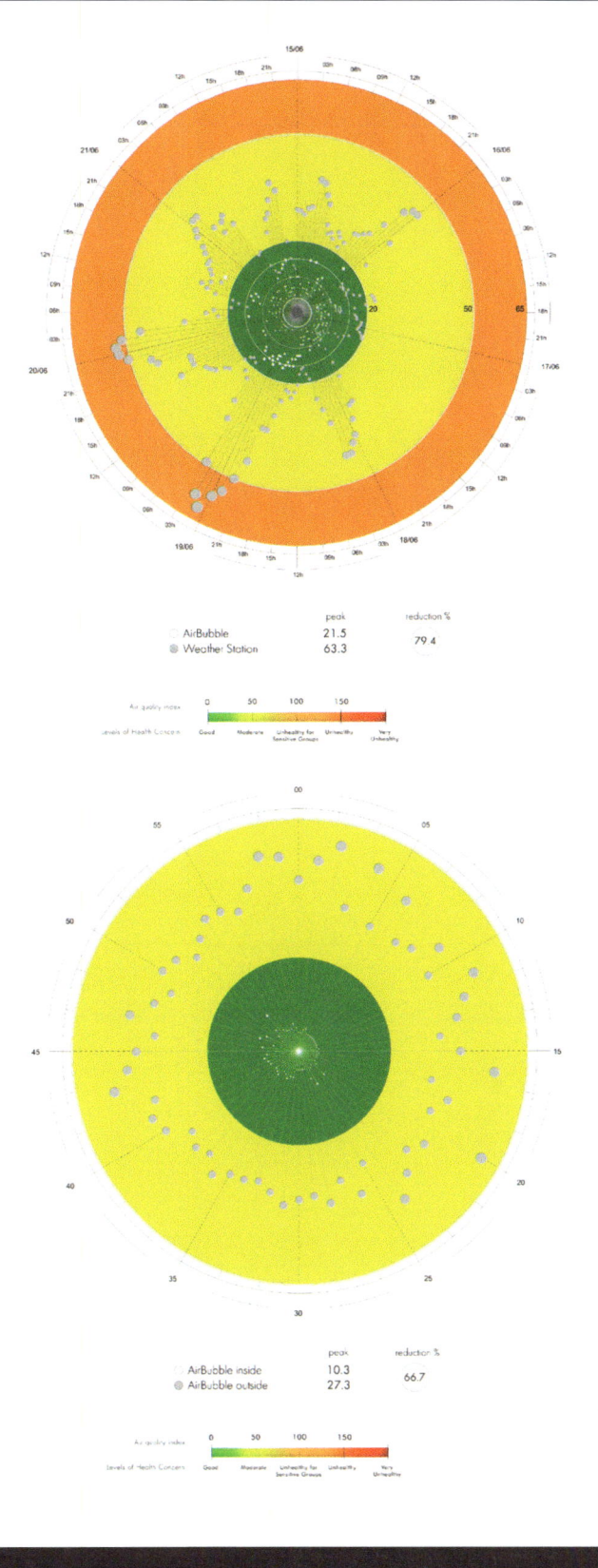

PM2.5 pollution monitoring diagrams for week 24 during summer 2021 in Warsaw. The two circular diagrams illustrate the comparison between concentration levels of fine particulate matter inside and around AirBubble. The top diagram shows the weekly comparison while the bottom diagram shows the peak hourly reduction.

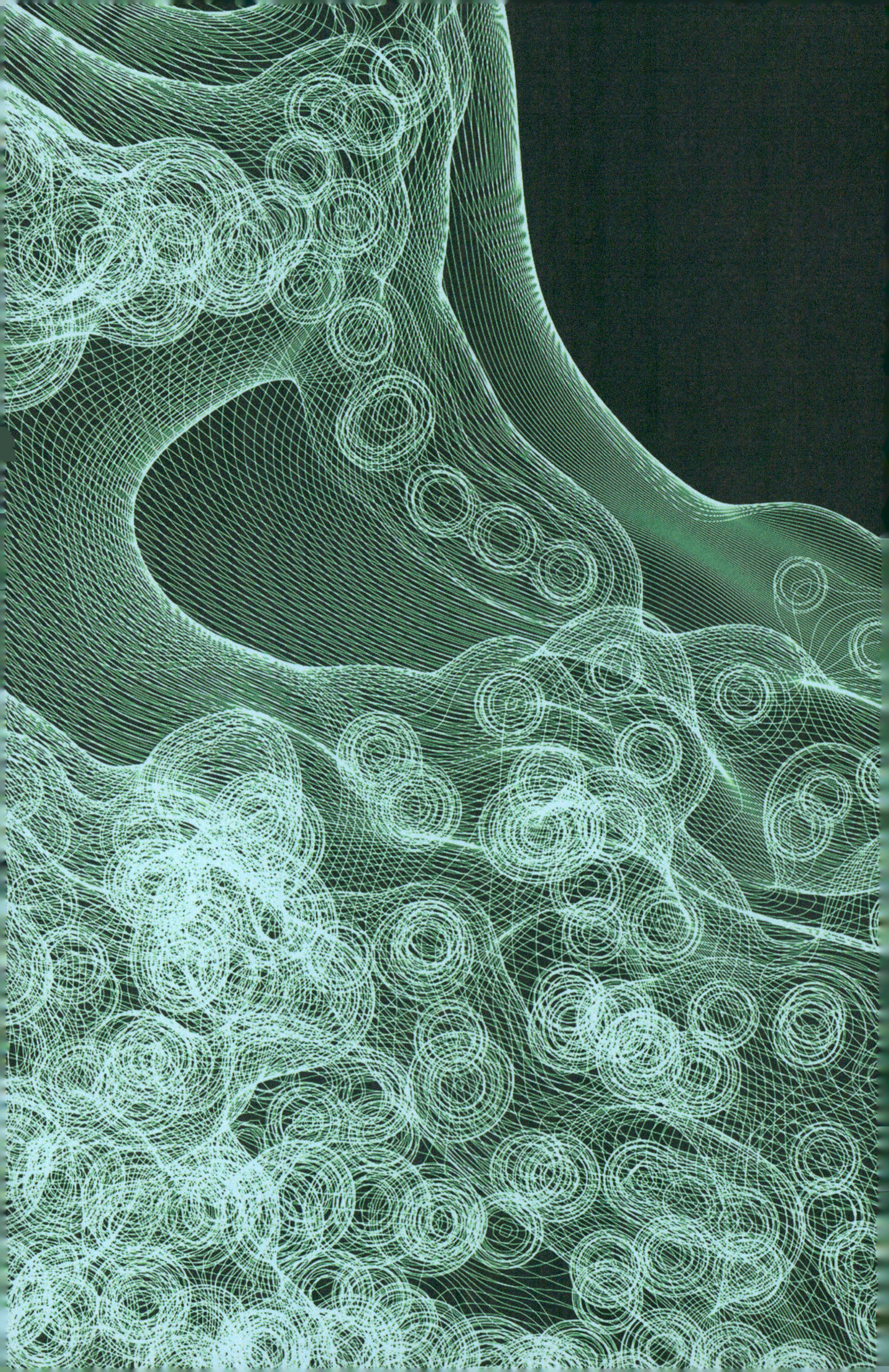

BIOTECHNOLOGICAL ARCHITECTURE: ON CULTURALISING THE URBAN MICROBIOME

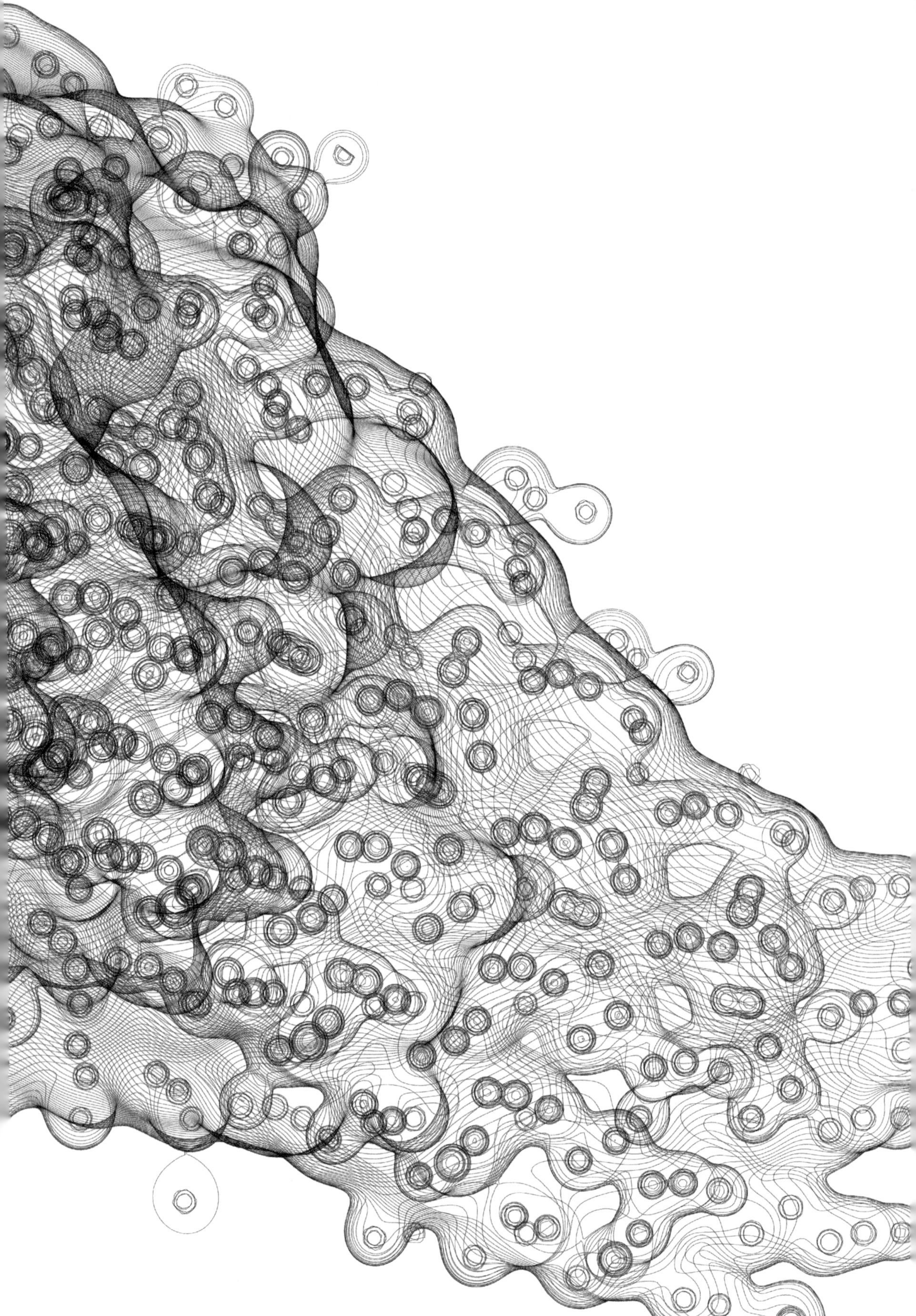

As data, molecules and cells become the materials of design, architecture is now defined by an evolving collection of spatial protocols of computation and practices of cultivation.

Today we live in the Anthropocene, the age in which human technology has come to affect virtually every ecosystem on Earth. Conversely, we are also in an era of ubiquitous computing, where the miniaturisation and distribution of digital systems has reached non-human levels of complexity and unpredictability. While our civilisation has impacted upon metabolic processes at the planetary scale, we have, perhaps paradoxically, become almost entirely dependent upon non-anthropocentric forms of intelligence. Human creations, such as urban infrastructures, now constitute new natures. The entire Biosphere is mutating into an Urbansphere, a stack of informational, material and energetic networks, which support our society's increasingly demanding metabolism.

We all inhabit the Urbansphere, which calls for architects to curate spatial knowledge across its domains. Liberated from the restraining boundaries of green ideology, the performative design practice described in this book reveals the beauty found in the abstract dimension of urbanspheric computation as well as in the most mundane aspects of urban infrastructure. As data, molecules and cells become materials of design, architecture is now defined by an evolving collection of spatial protocols of computation, of collective thinking, of practices of cultivation and of growing knowledge.

In order to provide the required levels of resources, in the right place and at the right time, the Urbansphere interrupts the fluctuating metabolisms of the other spheres of life on Earth. The miniaturisation, distribution and intelligence of the networks of the Urbansphere and of its nodes has reached non-human complexity, engendering evolving processes of synthetic life. Endosymbiotic relationships unexpectedly emerge among its heterogeneous components, especially when biological evolution negotiates contaminated habitats and ubiquitous forms of artificial intelligence. Therefore, in the Anthropocene, we seek to develop unprecedented strategies for dealing with such complexity and with the non-human logics that underpin it.

In this chapter, the authors discuss how such strategies can be developed within the realm of architectural and urban design through a combination of:

- direct observation of living organisms that operate collectively at scales other than the human scale;

- mediated interaction with related processes of material transformation and spatial morphogenesis; and

- radical repurposing of the bio- and digital technologies involved in these processes as tools of speculative design.

This chapter will also provide a cultural and architectural contextualisation of the design innovation presented in previous chapters, thereby showing the potential for its widespread adoption in the transformation of the built environment.

The Urbansphere and its technological apparatus, in the form of synthetic biology, biotechnology, artificial intelligence, and so on, opens scenarios in which the boundaries between natural and artificial, landscape and city, human and non-human realms are blurred. The object of architecture and the urban realm it inhabits becomes ambiguous, and its aesthetic language embodies feelings of estrangement, discomfort and disruption.

What emerges in the authors' practice can be defined as a productive form of alienation, suggesting that microorganisms such as bacteria, fungi, spiders and moulds can each act both as a behavioural model for design innovation and as active agents of architectural and urban production. This notion has thus far led ecoLogicStudio to experiment with processes of digital and biological computation, often embedding material intelligence in architectural apparatus at 1:1 scale. In the past 10 years the studio has built more than 20 prototypes, installations and pilot projects to describe, test and experience architecture as a form of material life. Some of the most significant examples are illustrated in the following pages.

These projects engage the evolving processes of living matter, thus embedding the objecthood of architecture within its surrounding environment or milieu. They highlight a new meaning for the notion of 'cultivation' that is no longer solely concerned with tending plants and natural landscapes, such as in horticulture, but is now involved in a more expanded field of analogue and digital design, impacting upon our perception and understanding of urbanity. As such, cultivation also acquires critical relevance in re-framing our relationship to emergent digital technologies and becomes part of a broader process in the 'culturalisation' of the non-human systems that populate the Urbansphere and which are now brought into the scope and focus of design and architecture in all of its aesthetic, performative and ecological aspects.

This shift has direct influence on the ever-critical relationship between form and performance in contemporary design. To unpack it, it is useful to adopt a bio-mimetic approach and observe the effects of endosymbiosis in collective formations like coral colonies. Critical to the definition of endosymbiosis and its relevance to design and architecture is the way that parts relate to wholes or, in other words, how we conceptualise the Urbansphere as a whole in relationship to its architectural and techno-material components.

Deleuze and Guattari,[1] in one of their best known passages, refer to the relationship between the wasp orchid and the thynnid wasp. The orchid flower has evolved parts that resemble very closely the female wasp; the seduced wasp male tries to mate with the flower and by doing so it pollinates the plant; the two have evolved so inseparably that even the appearance of the insect and plant have become similar. Despite them being obviously separate objects, belonging to separate realms of nature, the wasp is an essential part of the plant and of its reproductive mechanism—so much so that it becomes very hard to draw a frame around its identity. However, if we resist the temptation to categorise, we can instead conceptualise the pair as a larger ensemble and their coupling as reproductive process of a single machinic system. Similarly, a sand dune could be reconceptualised as the product of a vast array of sorting machines creating coherent patterns of sand distribution that travel in space and time until final dissolution. Even human beings might be better described as an assemblage of desiring machines, thousands of mechanisms which, unnoticed by us, are producing the dreams that we do notice as they surface into consciousness.

[1] Deleuze, G., & Guattari, F. (1987). *A Thousand Plateaus: Capitalism and Schizophrenia*. Minneapolis: University of Minnesota Press.

This notion of symbiosis forces us to redefine the modern and linear understanding of cause and effect by redescribing the boundaries of an object within its environment, and by also taking account of the multiple interlocking feedback loops that define the behaviour of the individual unit within the larger apparatus of the Urbansphere. In the evolution of some organisms the effect of such relationships has become so close that it has become internalised in the morphology and behaviour of the host organism itself, so much so that the organism appears to contain multiple levels of mechanistic feedback. Corals are a significant and fascinating case.

Corals are both collective organisms and also an example of endosymbiosis. An endosymbiont is a cell which lives inside another cell for mutual benefit. Eukaryotic cells are believed to have evolved from early prokaryotes that were engulfed in the digestive process of phagocytosis. If the engulfed prokaryotic cell remains undigested it contributes new functionality to the engulfing cell. In the case of corals this new functionality is photosynthesis. Over generations, the engulfed cell lost some of its independent utility and became a supplemental organelle. In the case of corals, the engulfed organisms are zooxanthellae, a form of dinoflagellate algae.

Corals are usually colonies of polyps. Polyps are alive coral tissue extensions that cover the calcium carbonate structure and are usually only a few millimetres thick. The tissue has two layers, the epidermis and the gastrodermis, where the zooxanthellae live. Zooxanthellae are unicellular and spherical with two flagella that fall off once they are incorporated within a host. The coral polyps do cellular respiration, thus producing carbon dioxide and water as by-products. The zooxanthellae then take up these by-products to carry out photosynthesis. The products of photosynthesis are used to make proteins and carbohydrates in order to produce calcium carbonate for the coral to grow their exoskeleton.

Polyps that have greater exposure to sunlight receive a competitive advantage via the zooxanthellae and are able to build their exoskeletons faster, thus gaining even more exposure to precious solar energy. Polyps that are less exposed and remain locked in will eventually die while the more exposed coral can reproduce and spread across larger areas. The particular morphogenesis of stony corals and their convoluted bifurcations is an emergent effect of this complex multi-layered process of symbiosis comprising multiple feedback loops between the marine environment, colony, individual polyps and the algae that inhabits their gastrodermis. It is therefore impossible to capture the nature of a coral's morphology without first understanding the dynamic nature of this rather complex part-to-whole assemblage and its purpose for survival and growth within a specific milieu.

The study of such complex interaction has fascinated scientists for decades, but it has only recently become possible to simulate digitally the emergence of this kind of morphogenesis, and thereby to study and visualise how this morphology emerges at a certain scale from multiple levels of interactions among parts that operate at a significantly smaller scale. One of the more interesting models is the so-called polyp-oriented model of coral growth, which argues that:

> "The morphogenesis of colonial stony corals is the result of the collective behaviour of many coral polyps depositing coral skeleton on top of the old skeleton on which they live. Yet, models of coral growth often consider the polyps as a single continuous surface. In the present work, the polyps are modelled individually. Each polyp takes up

resources, deposits skeleton, buds off new polyps and dies. In this polyp-oriented model, spontaneous branching occurs. We argue that branching is caused by a so called 'polyp fanning effect' by which polyps on a convex surface have a competitive advantage relative to polyps on a flat or concave surface".[2]

It is interesting to note here how the discretisation of the model, i.e. considering polyps as individual agents, affords a new understanding of the emerging nature of some key traits in the whole colony. This polyp-oriented model explains why morphologies that are species specific also show high variability within one species and how this intraspecific variability is caused by environmental parameters, such as light availability and the volume of water flow.

It also explains how the characteristic bifurcations emerge as a consequence of the collective nature of the coral. At convex sites, the polyps fan out, thus gaining better access to the diffusing resources. At concave sites, the polyps point towards each other, thus interfering in the uptake of resources. In this way, a curvature effect comes out as a natural consequence of the competition between the polyps to take up resources. It is possible to translate those findings into digital 3D meshes where polyps are vertices of the mesh with varying access to nutrients.

Two sets of considerations are particularly relevant here. Changing the nature of the environment (the amount of resources, their speed and direction of movement) without any change to the 3D coral mesh or to the rules of interaction of its polyps vertices does generate a very large number of different morphological traits in the coral mesh. Thus, survival in an environment of scarce resources produces diversity and richness of forms. Furthermore, the convolutedness of the resulting meshes reflects that found in actual coral morphologies and that results from the process of optimal solar exposure of a collective organism powered by photosynthesis. Small differences are amplified and lead to articulation of form. Thus, the complexity of articulation is a consequence of the economy of means.

Therefore, if we redefine our concept of part-to-whole relationships and trans-scalar hierarchies as being a continuum nesting heterogeneous systems composed of individually interacting units and immersed in a changing environment, we can come to understand performativity in terms of the generative forces of spatial and material articulation—morphogenetic force. Within this paradigm, competition for scarce resources leads to richness and diversity of forms. This is unlike most engineering problem-solving methodologies where optimisation leads to a convergence of 'average' solutions.

[2] Merks, R.M., Hoekstra, A.G., Kaandorp, J.A., & Sloot, P.M. (2004). Polyp oriented modelling of coral growth. *Journal of Theoretical Biology* 228 (4), 559–576. doi:10.1016/j.jtbi.2004.02.020

❝ This bio-digital perspective affords a new reading of the famous modernist credo of 'less is more'. Less resources equals more morphological articulation.

Over the last few years ecoLogicStudio's practice has become increasingly concerned with the actualisation of this morphogenetic paradigm in architectural and urban design. This is how the series of projects and installations known as H.O.R.T.U.S. was conceived and how the DeepGreen urban design paradigm evolved. Crucially, in these works the bio-mimetic model described above serves not merely as a metaphorical reference. The model of coral morphogenesis has been deployed as a generative design tool for architecture. It has evolved in an algorithmic design protocol and finally, as described in the previous chapter, it has inspired a fully fledged design innovation system. In other words, corals are what we might call 'bio-mimetic models': models which we deploy to design bio-philic artefacts that form a new landscape of man-made bio-digital living architectures.

Another important methodological consideration can be drawn from focusing our attention on part-to-whole relationships. This, once more, implies rejecting the metaphorical translation across scales and regimes to focus instead on the analogy between two models that may look and feel rather different but that share similar relational diagrams.

What interests the authors is that traits such as bifurcations and surface convolutedness maintain their emergent nature in relationship to a specific milieu even when translated into an architectural model. The intelligence of the model system can therefore be embedded in the here and now and it can operate across scales and regimes in the virtual and in the material realms simultaneously. The specific traits of the designed system are bred in time over successive iterations and throughout the design process, construction phase and period of usage. Next, we shall look at this important aspect in more detail.

3 Pasquero, C., & Poletto, M. (2012). *Systemic Architecture: Operating Manual for the Self-Organizing City.* London: Routledge.

In their book *Systemic Architecture*[3] the authors compared the methods of a bio-engineer synthesising artificial tissues in a laboratory with that of a gardener reviving a patch of dried-up earth: while both run generative protocols, the first requires a perfectly controlled testing ground for his procedures if he is to acquire generalised applicability, whereas the second needs to consider the unexpected fluctuations in the particular ecology of his garden. The gardener operates through a process of intensification of difference: his only chance for reconciling his desire for beautification with the natural expressivity of living processes resides in a form of movement—in both the biological and physical sense.

Architect and philosopher Gilles Clément suggests that, from this perspective, the formalisation of the garden becomes a process of formalised transmission of biological messages. Differences in slope, insulation, soil moisture and so on are registered and then exploited by the gardening protocol to promote the growth of different arboreal species. Furthermore, that growth, being itself a variable and partially unpredictable process, needs to be read, assessed and considered in the context of the formulation of future actions and in developing the lines of any future gardening protocols.

Gardens grow and beautification progresses in loops: each step generating more difference and local complexity that can in turn be recognised and bred. The management of this looping generative process is what makes the garden a potentially beautiful and healthy organism. In Clément's words:

> […] reality is entirely contained within experience. Only. Without gardening there is no garden.'[14]

This sensibility was already formalised in the realm of architecture by radical avant-garde visions of the 1960s, such as Archigram's 'Instant City'. The idea continued to evolve in the years that followed in the exquisite drawings of biomechanical landscapes and living architectures by Sir Peter Cook[5], which the authors consider to be a precursor of the new paradigm of the culturalisation of biotechnology. Peter Cook's aesthetic sensibility is ever more relevant today and is critical to the point we are making here. In fact, it prefigures the renewed continuity between the act of simulating (i.e. depicting) nature while at the same time hosting living systems—including of course, humans—within the ever-evolving fabric of architecture itself.

Architect Cedric Price was a contemporary of the Archigram group. He contributed what is probably the most enduring attempt to expand architecture into the realm of computation and embedded intelligence, especially in his long-standing collaboration with cybernetician Gordon Pask. Unbuilt projects such as 'Fun Palace' (1964) and the 'Generator' (1976), as well as the built Snowden Aviary at London Zoo (1962)—the latter is still in use today—present a 3D materialisation of the

5 Lindhardt Weiss, K. (2022). *Peter Cook on paper.* Copenhagen. The Danish Architectural Press.

paradigm of second order cybernetics and the related Paskian concept of environment. In the words of Prof. Ranulph Glanville:

"Second order Cybernetics presents a (new) paradigm—in which the observer is circularly (and intimately) involved with/connected to the observed. The observer is no longer neutral and detached, and what is considered is not the observed (as in the classical paradigm), but the observing system. [...] Therefore, second order Cybernetics must primarily be considered through the first person and with active verbs: the observer's inevitable presence acknowledged."[6]

6 Glanville, R. (2003). Second order cybernetics. In F. Parra-Luna (ed.), *Systems Science and Cybernetics*, vol. 3: *Cybernetics and the Theory of Knowledge*. Paris: UNESCO-EOLSS. www.eolss.net/Sample-Chapters/C02/E6-46-03-03.pdf

In architectural terms this idea suggests moving away from ideas of responsive or adaptive systems such as sun tracking louvres, so dear to the hi-tech modernists, to consider instead a participatory framework where the observer is no longer a mere user but becomes an active co-creator of the spaces he or she is inhabiting. This shift may enable a new kind of spatial conversation to emerge, one which may have previously unexpected consequences for the ways in which space is perceived and utilised, including opportunities for novel architectural forms. Architecture thus becomes a morphogenetic system that emerges from the meta-conversation between human and non-human systems. It is no longer a question of how we make use of the environment and the space we inhabit, but how we become co-creators of planetary space by interacting with other forms of intelligence.

Gordon Pask certainly understood the critical importance of circularity and its relevance to interaction and conversation theory. For that reason, he began building several devices that could interact with humans in space. These installations, such as the seminal 'The Colloquy of Mobiles' at ICA in London (1968), are known as 'Paskian Environments'. In Pask's own words:[7]

7 G. Pask cited in M.C. Bateson (2005 [1972]), *Our Own Metaphor*. Cresskill, NJ: Hampton Press.

"now we've got the notion of a machine with an underspecified goal, the system that evolves. This is a new notion, nothing like the notion of machines that was current in the Industrial Revolution, absolutely nothing like it. It is, if you like, a much more biological notion, maybe I'm wrong to call such a thing a machine; I gave that label to it because I like to realize things as artifacts, but you might not call the system a machine, you might call it something else."

♊ …we began calling them cyber-Gardens

One of the first cyber-Gardens ecoLogicStudio designed and built was StemCloud 2.0, a living bio-digital installation commissioned by the Seville Biennale of Architecture 2008 and precursor of the H.O.R.T.U.S. series. Designed in the form of a coral-like landscape, it invited the public to climb up close and interact with its bio-reactors. Each hosting living microbial cultures, they were stacked into a curved formation which framed the gallery space, creating an accessible niche, screening light and releasing oxygen into the atmosphere. Itself a collective organism, the installation behaved as a self-regulating photosynthetic body where both 'external conditions', so-called 'environmental factors', as well as the particular micro-ecologies contained in the bio-reactors become integral parts of the cyber-gardening process. The project is described in detail in the book *Systemic Architecture*.

Almost ten years later, this line of research has evolved into ecoLogicStudio's first permanent bio-digital sculpture, 'H.O.R.T.U.S. Astana', now part of the Bio.Tech Hut at the Museum of Future Energy in Kazakhstan. Its morphology embodies the three-dimensional distribution of wide spectrum lighting sources. The flow of energy emitted as wide spectrum radiation is digitally simulated in space to visualise its intensive field. Cyanobacteria are then introduced as bio-bits, their metabolic machines deployed to convert radiation into actual processes of photosynthesis, oxygenation and biomass. Their articulation in space is digitally mediated to arrange the photosynthetic organisms along iso-surfaces of equal incoming radiation. A network of connecting paths is also computed, bringing nutrients and CO_2 to the living cyanobacteria. Visitors are active participants in the system, feeding it CO_2 with their exhalations and absorbing the released oxygen as it spreads in the surrounding atmosphere.

The project recasts the practice of computational design into cyber-gardening, embedding biological computation in space. This project represents an attempt to culturalise the non-human. Or, in other words, to re-describe the boundaries of generative practices beyond what we have called the 'Lab model' and towards extended bio-digital cultivation, the 'cyber-Garden model'. This model has reached its most advanced definition in the living sculpture 'H.O.R.T.U.S. XL', commissioned by the Centre Pompidou in Paris in 2018. In H.O.R.T.U.S. XL the digital algorithm simulates the growth of the exoskeleton that is then deposited by digital 3D-printing machines in layers of 400 microns, triangular units of 46mm and hexagonal blocks of 18.5cm. Photosynthetic cyanobacteria in a bio-gel suspension are then introduced into the individual triangular cells to the form units of biological intelligence of the collective system.

In H.O.R.T.U.S. XL, intensive difference is artificially created, and the living cultures are forced to negotiate these differing gradients, thus evolving novel forms of biological intelligence visible as unique morphological traits. In this sense, the project casts the significance of urban ecology in a new light. It promotes the emergence of what we have called 'non-human survival strategies'. Novel behaviours emerge from the coexistence of man-made artefacts, machining production protocols and living biological organisms. The relevance of such behaviours to the production of architecture must be evaluated in relation to the conception of the city as a living Urbansphere.

In fact, while much has been said about the crucial transition between the wasteful mass production of artefacts and towards a digitally enabled mass customisation of processes, we would argue that very little has been accomplished in respect of investigating the ways that digital production of space can enable the proliferation of novel models of inhabitation or forms of spatial intelligence. Biological life proliferates from exploiting the potentials of intensive difference. It is constantly opportunistic and open to the actualisation of new material systems.

From this perspective, the authors argue that in order to evolve a resilient Urbansphere we should intensify its mechanisms of spatial intelligence while also rescuing them from the conforming forces of contemporary technological innovation that are clearly recognisable in the dominant paradigms of both 'smart' and 'green' cities.

The field of design innovation shall embody a sensibility that some researchers are now loosely defining as post-digital but that in thought has significantly older roots. In the words of Félix Guattari:[8]

8 Guattari, F. (2008). *The Three Ecologies.* London: Continuum.

> "So, wherever we turn, there is the same nagging paradox: on the one hand, the continuous development of new techno-scientific means to potentially resolve the dominant ecological issues, [...] on the other hand the inability of organized social forces and constituted subjective formations to take hold of these resources in order to make them work."

Embedding technology into autonomous forms of material organisation augments architecture as a spatial interface involved in everyday ecological practices of cultivation — that is the culturalisation of the Urbansphere. Architectural technologies can thus operate at multiple interconnected scales at once, from the micro-regimes of algae cells to the macro-behaviours of urban infrastructures, expanding architecture in both scope and reach. It provides the spatial substratum required by extended participation, which we have named 'bio-citizenship', underlining the expanded role played by non-human entities in future collective formations.

In the future, every architectural protocol should seek to enable extended participation and, as a consequence, architecture must be able to grow in complexity and itself embody a form of artificial intelligence, one that is distributed in the fabric of the spaces we inhabit and that is co-evolutionary with them. In this respect, microalgae are to be understood not only as biological organisms, as found for instance in their symbiotic relationship with polyps that influences coral morphogenesis, but also as a fascinating repository of intelligent survival strategies that has evolved over millions of years and that can be embedded into future biotechnological architectures such as ecoLogicStudio's proposed 'PhotoSynthetica Tower'.

At the city scale, the Tower appears as a complex synthetic organism in which bacteria, autonomous farming machines and other forms of animal intelligence become bio-citizens. Bio-digital research units, gardening centres, wildlife observation terraces, self-sufficient dwellings and a potentially infinite variety of other programmatic combinations are supported by the continuous catalytic action of the Tower, which constantly re-metabolises anthropic pollution and biotic contamination into local, circular economies of raw materials, data and energy.

In the PhotoSynthetica Tower, as well as in ecoLogicStudio's test beds in Dublin and Helsinki illustrated in the previous chapter, the contribution of human and non-human stakeholders is nurtured by elevating both to the role of co-creators of the actual solutions in their becoming. In this sense, it is less important whether the design medium includes algae, bacteria or other kinds of digital systems. All these forms of intelligence are involved in the creative process. Designers, eventually, can release control on the actual definition of the final architectural artefact and enjoy engaging in its evolutionary process. Crucially, the creative practice simultaneously gains in scope and in the ability to influence the morphogenetic process at a more profound level.

A protracted experimentation with biological and digital design technologies for the urban realm surely enables the development of innovative solutions to the urgent ecological problems affecting our cities. However, it is not sufficient to operate technologically. That is why ecoLogicStudio in its professional practice, is intensively testing a new generation of bio-digital design prototypes with transformative agency for an architectural discipline yet to come.

The Tower represents a future biotechnological architecture that is plural, collective and mutant, defined by the trans-scalar nesting of heterogeneous systems each composed of individually interacting units immersed in a changing environment. Within this new paradigm performance is a generative force of spatial and material articulation, a morphogenetic force. Adaptation to a world of scarce resources can thus lead to a richness and diversity of forms and to increased material intelligence.

This is a process that has no beginning and no end. Rather, it evolves in time from conception to construction and beyond, and throughout its useful life cycle—suggesting a shift away from ideas of responsive or adaptive environments and toward developing participatory frameworks where the observer is no longer a mere user, but becomes an active co-creator of the spaces he or she is inhabiting. Ultimately, computational design strategies could be re-cast into a broader practice of embedding bio-digital intelligence into physical space. This project represents an attempt to 'culturalise the non-human', that is, to re-describe the boundaries of generative design practices and related research beyond the constraints of wholly human rationality and creativity.

At this point, it is worth considering a project called 'meta-Folly', which ecoLogicStudio completed in 2012. It is a small pavilion, now part of the permanent collection of the FRAC Centre in Orléans. It is alive but does not involve or contain any living biological material. In 2018, as part of a large show at the CCCB in Barcelona dealing with the topic of sensuality in architecture, meta-Folly was exhibited as a connecting piece between original drawings of Archigram's Instant City and photos of the famous house prototype Villa Rosa (1968) by Coop Himmelblau. The curators raised the question of how we might treat digital materialism as a cultural and social critique, and of how such critical, post-digital attitudes could serve the practice of architecture in terms of fostering its relevance at the cutting edge of social and ecological innovation. This perspective reveals an important interpretation of the role of the creative practice, that of being a form of research.

In the words of Claudia Pasquero and Emmanouil Zaroukas.[9]

9 Pasquero C., & Zaroukas, E. (2016). Design prototype. *AAE Conference Proceedings*. London: The Bartlett UCL, pp.96–108.

"To question the constitution of problems and to turn that question into an affirmative proposal. [...] innovation is effected by experimenting with the problematic field and not exclusively with the solution space. With problematics we identify the domain through which problems are formed. Design in this case then becomes the process of designation of a problem and the production of knowledge is effected by re-framing the problematic. In this sense, not only history reads differently, but also problems are constituted differentially."

Architecture thus begins to operate at multiple interconnected scales simultaneously, from the micro-regimes of algae cells and pollution particles, to the macro-behaviours of smart urban infrastructures, expanding both its scope and reach. It provides the spatial substratum for re-problematising problems and recognising the expanded role played by non-human entities in the Anthropocene age.

In the future, architectural protocols should seek to enable this extended participation and to embody its artificial intelligence—a form of intelligence that is distributed into the fabric of the spaces we inhabit and that is co-evolutionary with them.

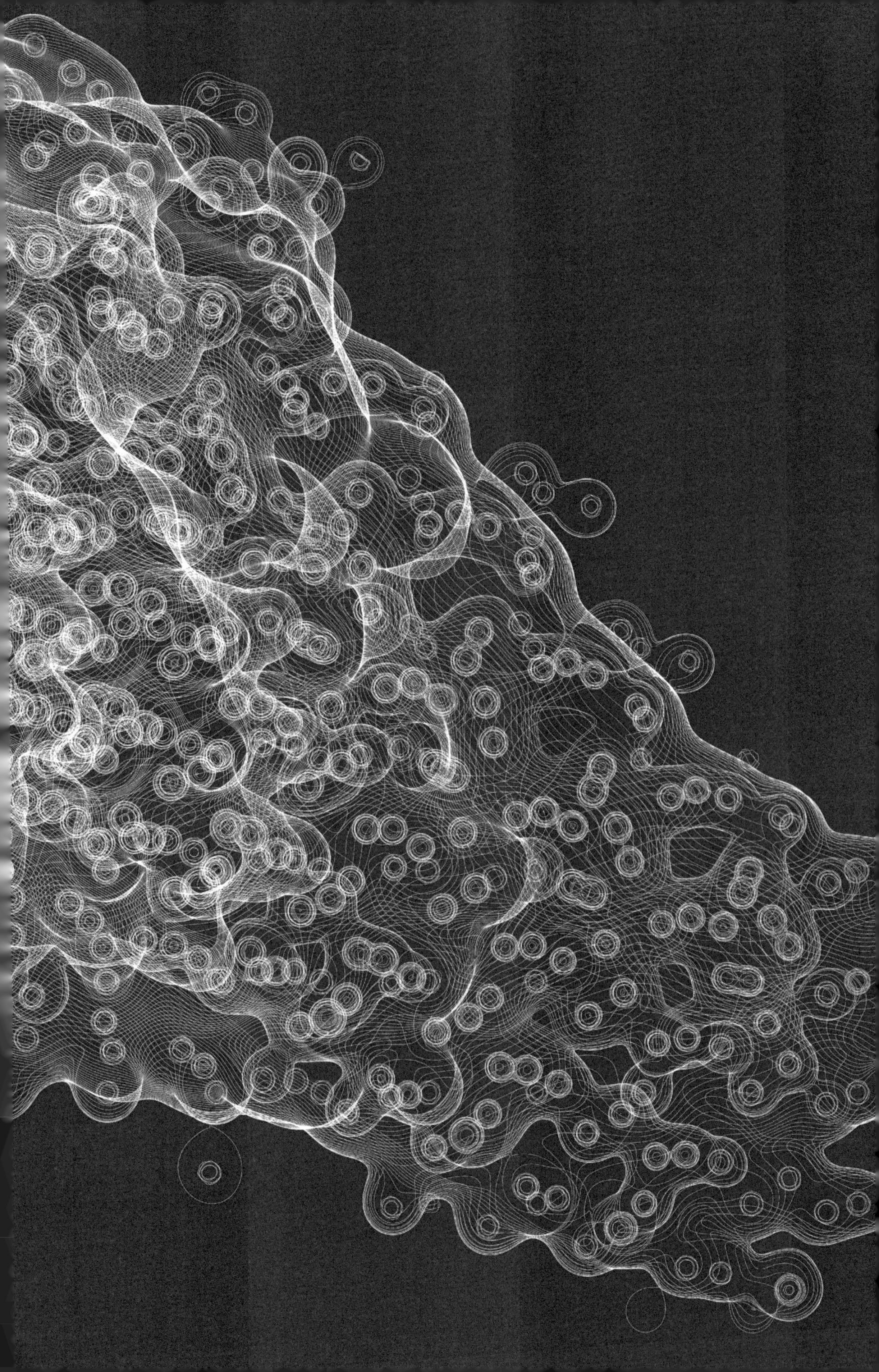

Projects 1.02

BIO-DIGITAL SCULPTURES: DESIGNING THE LIVING

1.02.01 SuperTree 128

1.02.02 H.O.R.T.U.S. XL and the PhotoSynthetica Tower 136

1.02.03 Arbor 150

1.02.04 bI.O.serie 158

1.02.05 meta-Folly 162

SuperTree is an architectural apparatus that repurposes the archetype of the tree to optimise its infrastructural operations as part of the present Urbansphere.

1.02.01

SuperTree
Permanent collection
ZKM, Karlsruhe

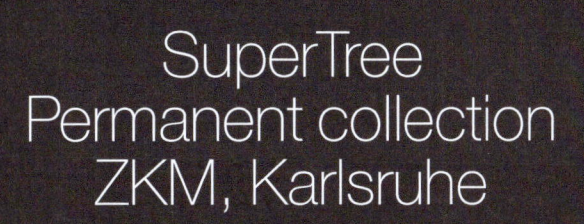

The SuperTree is a living bio-digital sculpture conceived and produced by ecoLogicstudio and is part of the permanent collection of the ZKM Centre for Media Art in Karlsruhe, Germany. It was last exhibited on the occasion of the opening of the new Futurium museum in Berlin.

Trees, the most potent symbols of our traditional ideal of nature, have now evolved into carbon sinking machines, shading devices, flood barriers, building materials, marketing tools and so on, serving the Urbansphere's increasingly demanding metabolism while also providing a large part of its soft interface with other living creatures, including us. This is a challenging role, one that requires radical adaptation capabilities and that is often incompatible with the slow pace of biological evolution. SuperTree transforms the archetype of the tree into a high-resolution, high-productivity photo-bioreactor apparatus that connects human metabolism to the proliferation of life within microalgal ecologies such as cyanobacteria cultures.

Flows of energy (light radiation), matter (proteins, CO_2, cyanobacteria) and information (morphology, DNA) are intercepted, processed and returned by SuperTree with efficiency that is one order of magnitude greater than a conventional tree. Current advances in biotechnology and digital design and fabrication are so rapid that this could increase ten-fold every 10 years, enabling rates of progress in the overall resilience of the Urbansphere that were previously inconceivable.

Cover page: Visitors interacting with the cyanobacteria cultures in the fruiting bodies of the SuperTree at Futurium, Berlin.

This page: SuperTree exhibited at the inauguration of Futurium, Berlin.

Close-up view of the canopy with fruiting bodies and cannulae for human–algae interaction.

The core structure of SuperTree is algorithmic: light radiation iso-surfaces are computed to determine the boundaries of the canopy; Voronoi tessellation is deployed to develop its porous internal structure; and the branching logic defines its trunk and the rooting system at its base. Laser-cut aluminium makes it extremely lightweight, reflective and porous to incoming solar or wide spectrum light radiation. A new version in robotically printed algae-based biopolymers is already being produced and will achieve a higher resolution and incredible mechanical resistance while also being completely biodegradable.

The canopy and main trunk are entirely photosynthetic, powered by colonies of cyanobacteria which produce ten times more oxygen than the leaves of a tree of similar size. By introducing the selective breeding of the microalga Cyanidium, SuperTree achieves its robust behaviour even in the most challenging urban environments, including wide temperature fluctuations and exposure to highly polluted atmospheres.

In SuperTree, the microorganisms flow within a 6km long hose, thereby maximising their exposure to radiation. CO_2–O_2 exchange occurs at the base and at the fruiting bodies.

These hanging reactors host tendrils for human–plant interaction. Humans exhale CO_2, thus feeding the bacteria the carbon they need to build biomass. In exchange, O_2 is released for humans to breathe. 60% in mass of wet algae is vegetable protein oils, and so as biomass builds up within the fruiting bodies their protein content increases exponentially. This protein mass can then be drawn directly out of the fruiting bodies via silicon cannulae.

To close the carbon cycle, SuperTree can generate biomass for bio-digestion by other urban bacteria colonies, emitting bio-gas and soil fertiliser as a result.

Detail of the photosynthetic iso-lines along the SuperTree canopy.

With H.O.R.T.U.S. XL Astaxanthin.g we have designed an architecture that is receptive to microbial life.

1.02.02

H.O.R.T.U.S. XL and the PhotoSynthetica Tower

H.O.R.T.U.S. XL Astaxanthin.g is a large-scale, high-resolution 3D-printed bio-sculpture receptive to both human and non-human life. The project, first commissioned by the Centre Pompidou in Paris, was conceived by Claudia Pasquero and Marco Poletto (ecoLogicStudio) and developed in collaboration with the Synthetic Landscape Lab at the University of Innsbruck.

H.O.R.T.U.S. XL was presented for the first time at the Centre Pompidou in Paris as part of the exhibition 'La Fabrique du Vivant' which opened in Paris in February 2019. Subsequently, it travelled to Vienna at the MAK for the Architecture Biennale, to the Mori Art Museum in Tokyo, to the Fundation Telefonica in Madrid and to Hyundai Motor Studio in Busan, South Korea. Recently, a new version of H.O.R.T.U.S. XL has been acquired by the Neom project in Riyadh, Saudi Arabia.

In the digital age, a new interaction is emerging between creativity and the fields of life science, neuroscience and synthetic biology. The idea of 'living' takes on a new form of artificiality. This project confronts the dictates of human rationality with the effects of proximity to bio-artificial intelligence. It is envisioned as a collaboration with living organisms, their non-human agency mediated as it were by spatial substructures developed from biological models of endosymbiosis.

In particular, this sculpture is inspired by studies conducted by the authors on the collective behaviour of coral colonies and their morphogenesis. Individual coral polyps host microalgae called zooxanthellae within their tissues. As the algae photosynthesises, it provides a metabolic flow of energy to the polyps, which in turn build their exoskeleton of calcium carbonate.

H.O.R.T.U.S. XL Astaxanthin.g deploys an algorithm to simulate the growth of a 3D substratum inspired by coral morphogenesis. The result is a set of digital meshes that are subsequently analysed and selected as inner and outer layers of the 3D-printed substratum of the sculpture. Each vertex of the mesh represents a virtual coral polyp. The substratum is further developed into a 3D-printable structure. This structure, as in the case of corals, is developed to support the proliferation of colonies of cyanobacteria that will inhabit its individual cells (bio-pixels). Each cell is therefore occupying the interstitial space between inner and outer layer. These two layers are then translated into a porous field of contour lines indexical of incoming solar radiation. The curvilinear profiles provide partial enclosure to the cells while enabling light penetration and oxygen–CO2 exchange.

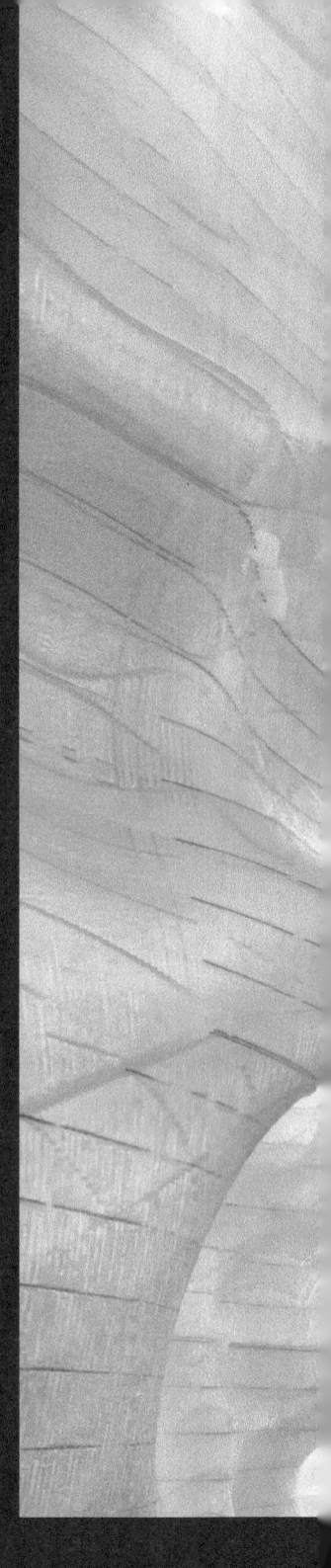

Cover page: Inoculating living Chlorella cultures on H.O.R.T.U.S. XL at Centre Pompidou in Paris.

This page: Close-up view on the interior of H.O.R.T.U.S. XL, highlighting its biophilic architectural skin hosting 'bio-pixels' of living Chlorella.

Close-up view of the inoculation process of living Chlorella cultures into bio-gel-based medium.

H.O.R.T.U.S. XL before the opening of the Fabric of Living exhibition at the Centre Pompidou in Paris, France.

H.O.R.T.U.S. XL at the Mori Art Museum in Tokyo, Japan. In the background the skyline of Tokyo can be seen. Photo: Kioku Keizo.

The final digital model of the substratum is then prepared for 3D printing in PETG (polyethylene terephthalate) on a Wasp 3D-printing machine. The layering process is algorithmically controlled to match the curvilineal profiles of the outer layers with the actual tool paths of the 3D-printing nozzle. Each layer is 400 microns thick with triangular infill units of 46mm. It is printed in 105 hexagonal blocks of 18.5cm each side producing an overall substratum that reaches 317cm in its tallest point.

Photosynthetic cyanobacteria cultures are then inoculated in a bio-gel medium and into the individual triangular cells, or bio-pixels, forming the units of biological intelligence of the system. Their metabolisms, powered by photosynthesis, convert radiation into actual oxygen and biomass. The density value of each bio-pixel is digitally computed in order to optimally arrange the photosynthetic organisms along surfaces of progressively higher incoming radiation.

The latest innovations in digital design and 3D printing thus enable cyano-bacteria's unique biological intelligence to become an active agent of architectural innovation. The scales of architectural detailing and the urban microbiome thus become compatible for the first time in history, conjuring up a new form of bio-digital architecture. Noticeably this enables multiple interactions in buildings that can now be activated by the intelligence of microalgae colonies, promoting a new kind of bio-architectural symbiosis.

This architectural symbiosis was explored more in depth in an exhibition presented at the Mori Art Museum in Tokyo in November 2019. Suspended at the 53rd floor of the Mori Tower and with the backdrop of Tokyo's urban sprawl, H.O.R.T.U.S. XL materialises its urban dimension in the form of a new prototype for living architecture, the PhotoSynthetica Tower.

At the city scale, the Tower is a large synthetic organism in which bacteria, autonomous farming machines and other forms of animal intelligence become bio-citizens. Alongside humans they all contribute to the new formation of Tokyo's own synthetic urban landscape. The combination of the intricate morphology of the PhotoSynthetica Tower with its sheer scale promote a significant microclimatic effect. The prevailing winds generate enough draught and turbulence to force both natural seeds and air-polluting particles through its porous skin. Each module of this skin is then activated locally to enable it to evolve its unique function.

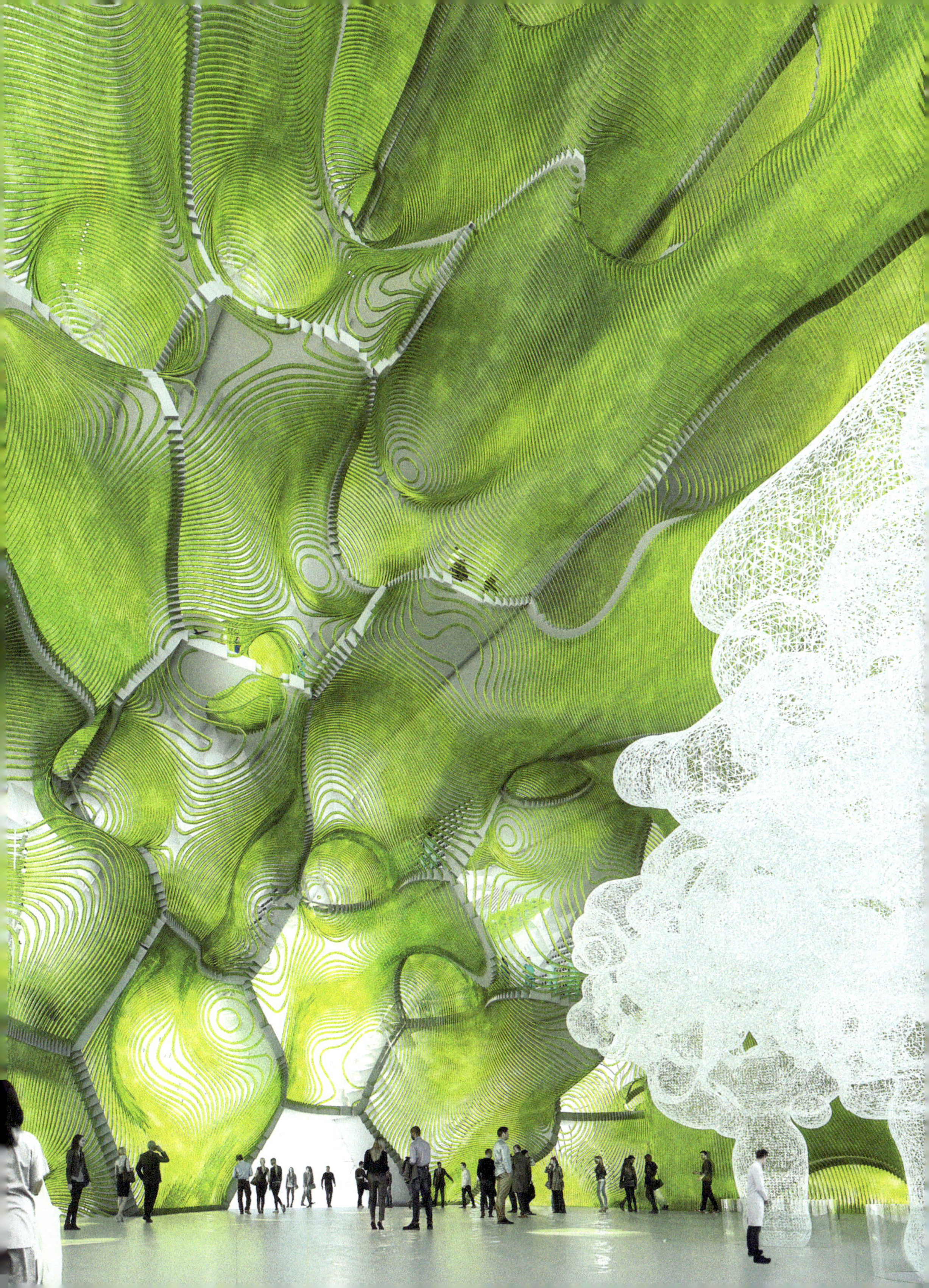

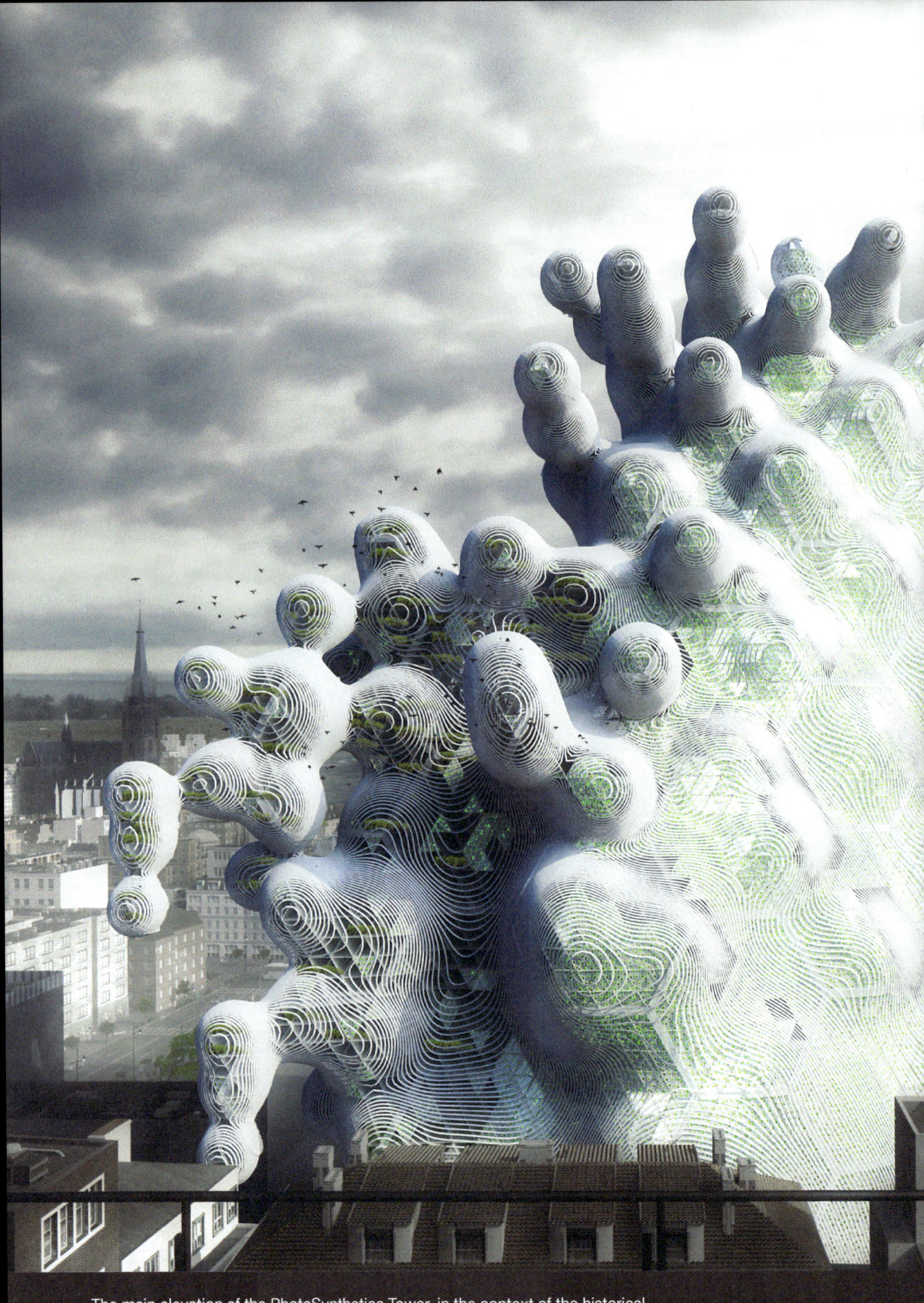

The main elevation of the PhotoSynthetica Tower, in the context of the historical centre of the Austrian city of Linz.

Some components of the system are designed as photo-bio-reactors. These are custom-printed bioplastic containers that focus sunlight to feed living microalgal cultures and release luminescent shades at night. Unfiltered urban air is pushed through the bottom of each module causing air bubbles to rise up naturally through the liquid medium within each photo-bioreactor, thus coming into direct contact with the voracious microbes. CO2 molecules and air pollutants are then captured and stored by the algae and grow into new biomass. Freshly photosynthesised oxygen is released at the other end of the module and channelled into the vast inner lobby of the Tower. In this way a clean urban microclimate is synthesised, supporting air circulation within a shared public space.

Other components become receptive to seeds and wild plants, thus forming emerging artificial habitats. These biotopes remain open to wildlife, including insects and migrating birds. The biomass that grows on all the active areas of the Tower is made available to the occupants of the building itself, supplying a plethora of emerging activities that define the programmatic mix of the building and its occupational patterns, for both human and non-human inhabitants. Bio-digital research units, such as Bio.Tech Huts and AirLabs, gardening centres, wildlife observation terraces, self-sufficient dwellings and a potentially infinite variety of other programmatic combinations, are all supported by the continuous catalytic action of the Tower as it constantly re-metabolises anthropic pollution and biotic contamination into local circular economies of raw materials, data and energy.

Close-up view of the PhotoSynthetica Tower main elevation highlighting the air-purifying cladding and sky habitats for wildlife.

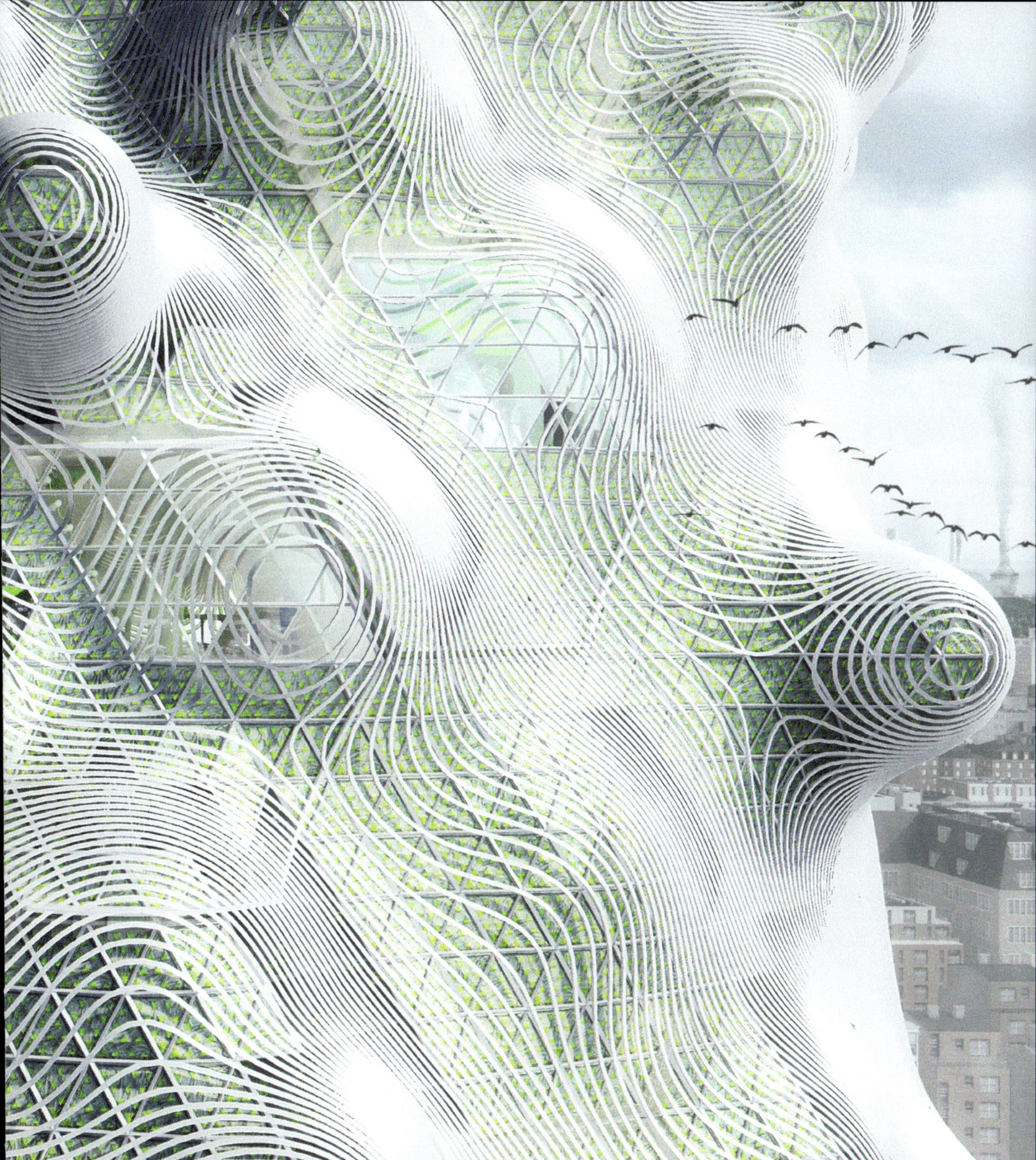

Arbor rethinks the life cycle of wood, proposing a bio-artificial system which is alive in a cybernetic sense.

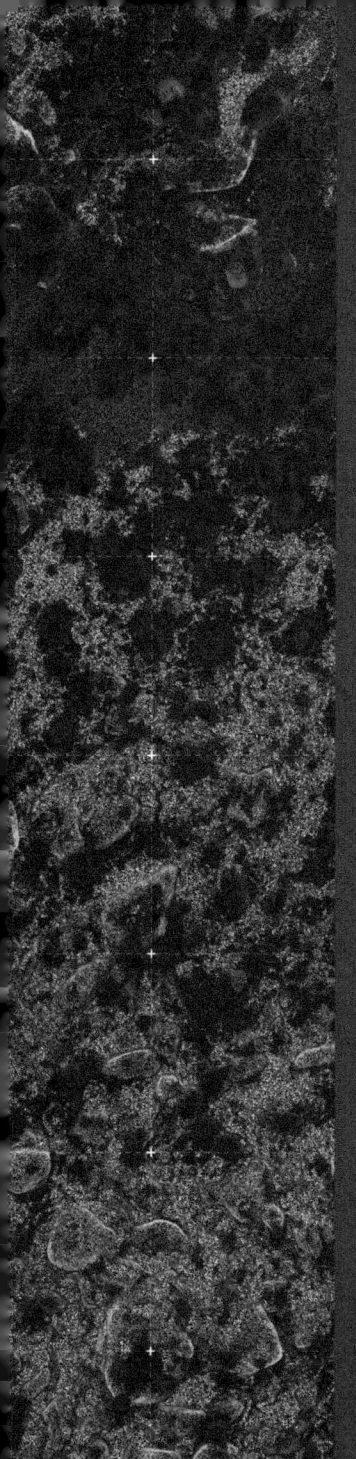

1.02.03

Arbor

Maria Kuptsova

If we look at the microscopical pattern of wood we see a very high degree of complexity embedded in its structure. It contains information about intelligent mechanisms of photosynthesis, growth, water and food distribution. According to Vincent, Bogatyreva, et al.:

> "Biological structures adapt to external stimuli by growth-induced material property variation resulting in hierarchically structured forms. Shape results from bottom-up material organization, sophisticated property gradation and functional hierarchies developed over time within single material systems." (Vincent, Bogatyreva, 2006)

By embedding the organisation principles of an organic material into a digital system we can design a hybrid materiality that hosts biological intelligence within a digital structure. The project 'Arbor' outlines an approach for reading the intelligence of an organic timber structure by means of machine-learning algorithms, and rethinks the life cycle of wood, proposing a bio-artificial system that is alive in a cybernetic sense.

Developed within the framework of doctoral research by Maria Kuptsova and under the supervision of Professor Claudia Pasquero at the Synthetic Landscape Lab, this project proposes the use of generative adversarial networks (GANs) as a method for extrapolating a material organisational principle from an existing database of timber structures.

Wood is widely used in architectural construction because of its mechanical, environmental, economic, acoustic, elastic, thermal, hygroscopic and aesthetic properties. However, the heterogeneous properties possessed by wood in its living state may also be problematic. According to Dinwoodie:

> "Unlike other construction materials developed to meet specific manmade functional requirements, wood has evolved as a highly efficient biological system—a vascular tissue—to meet the support, conduction and storage requirements of trees." (Dinwoodie, 2000)

The Arbor project uses the large database of the botanical characteristics of different wood species developed by the ArchiWood project as a resource for study of the material organisation of wood. Micro-images of 995 species were collected and analysed using various techniques. The anisotropic properties are crucial characteristics of timber structure which represents its constant and reciprocal interaction with the surrounding environment. For example:

> "For each wood species, there are 3 anatomical cuts associated with the 3 planes of symmetry of the material taken in 3 different magnifications of the microscope: x40 (overview of the cut), x200 (main features anatomical from IAWA), x400 (highlighting fine lines such as punctuations)." (Hallé, Détienne, Corbière, 2017)

Many of these qualities, which are often considered to be negative characteristics for traditional construction purposes, may be rethought and reused in the design of artificially alive systems that are capable of interacting with the environment and may develop additional functional capacities through this interaction.

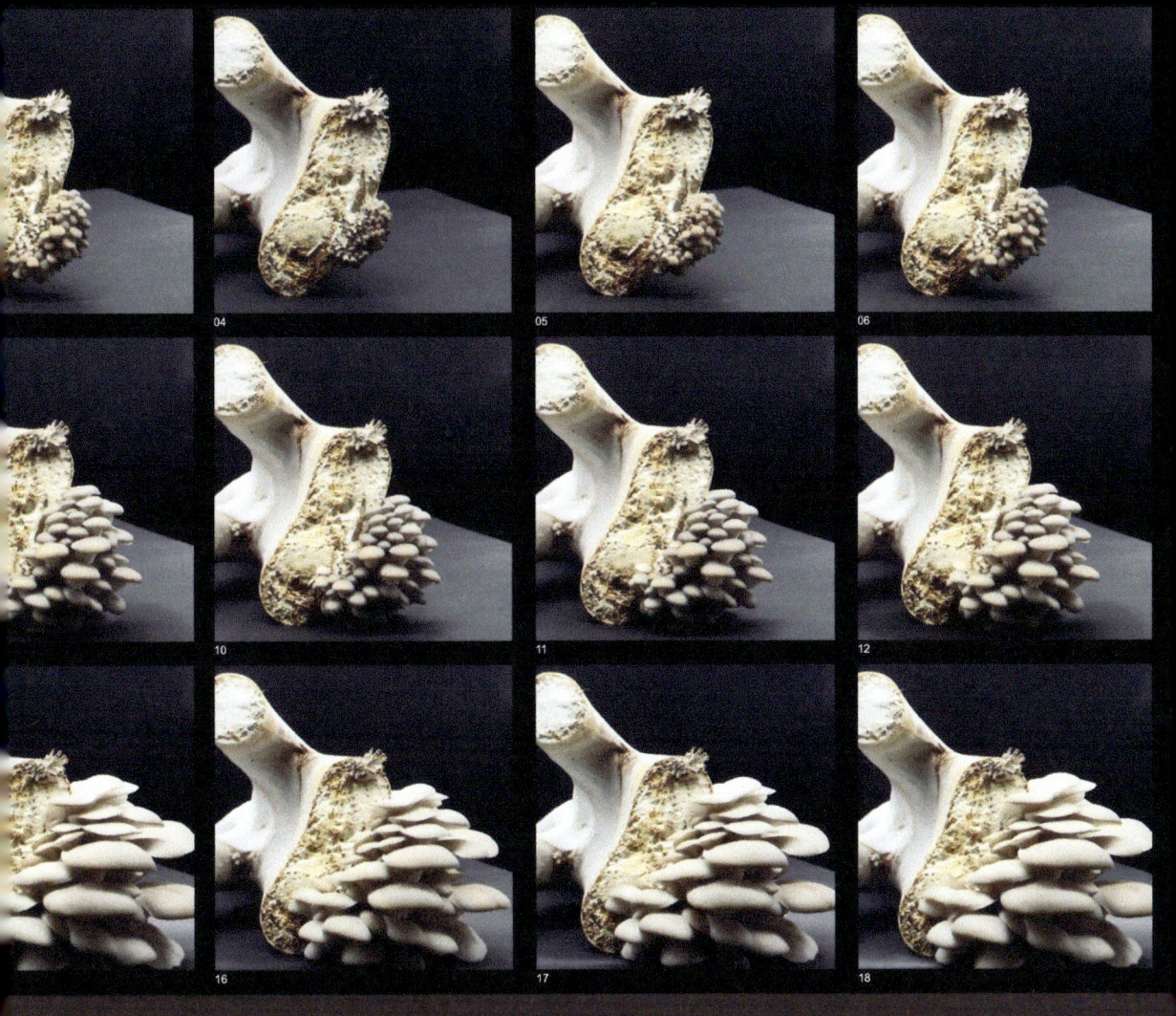

Cover page: Aerial 3D scanning in alpine region, photogrammetry. Image by Matthias Vinatzer.

This page: Material system for re-metabolisation of plastic waste: oyster mushroom, 3D-printed substrate. Image by Charles Wirion.

Close-up view of the fine 3D-printed timber structure highlighting hybrid materiality.

For the Arbor design research and in order to develop a technique that would allow organic data to inform inorganic computational systems, the StyleGAN2 variant of GAN was chosen for its specific evolutionary neural network algorithm. A set of experiments was used to train the algorithm on all types of wood cuts and magnifications. Eventually, the transverse cut in the magnification x200 model was selected as it was found to be the most consistent.

This x200 model was used to study the internal and external morphology of timber material organisation with the latent-vector interpolation translation into the Z axis of volumetric structure. In this process, basic geometrical data describing the anatomical properties of the wood structure of a thousand of timber species were extracted and memorised within a spatial topology. The discontinuous volumetric data-set describes several aspects of the material organisational principles: allocation of stiff and soft materials within a structure; gradients of fibre densities; and variation in hydrophilic properties. These networks of curves and surfaces describe the distribution of material as behavioural patterns.

Fabrication technologies such as additive manufacturing allow for the development of adaptive fabrication methods informed by research in material behaviour. A wood-based material system has been investigated through large-scale 3D-printing experiments, suggesting a new regenerative life cycle of matter from the wood in its living state through to the recycled wooden material.

Two different material systems have been tested for the development of the synthetic prototype. The first used wood-based filaments containing 40% wood and 60% recycled PLA plastic. The second was focused on the development of a material system based on 99% content of wood with 1% agar agar (a jelly-like substance derived from red algae) as a binding material. For this test a custom-made end-effector was designed to control the accretion of wood paste. Material studies like these can provide fresh insight into how additive manufacturing can be used to develop novel approaches for timber construction based on bio-printing. Both methods have shown successful results and have been tested at a large scale. For the production of the Arbor sculpture, the wood-based filament with 40% content of wood powder was chosen as it was more suitable for large-scale rapid prototyping.

In combination, the use of robotic manufacturing, the regenerative life cycles of matter and the introduction of machine-learning-based design techniques could potentially support the artificial growth of cyber-organic wooden structures by means of intelligent technologies. In this way, the established understanding and application of processes of growth, decay and ontogenesis may be challenged by introducing a form of cyber-organic living object, bio-artificially grown in the Urbansphere.

According to Pask, the growing field of computing and synthetic biology, humans, non-humans and their shared environment "might coexist in a mutually constructive relationship" (Pask, 1976). Architecture and urban design play a role in enabling new frameworks of communication and co-living between human and non-human. The potential for imparting biological intelligence into inorganic systems creates a new generation of design objects which "are no longer limited to assemblages of discrete parts with homogeneous properties, like in modular systems, or to continuous gradient fields of material articulation, as in parametricism" (Hui, 2019). Instead, design objects are bio-artificially grown, forming materially heterogeneous systems with a complex functionality and morphology.

The system we are describing is a regenerative system in which the overall organism is alive and the concept of life involves both biological and technological forms. The intelligences of biological, geological, ecological, algorithmic and technological systems are an active agent of design. This bio- and "techno-diverse" design system forms novel spatial topologies and aesthetic constructions for an advanced model of the cyber-organic city (Orive, Taebnia, Dolatshahi-Pirouz, 2019). The cyber-organic city provides the conditions for an inclusive democratic multispecies polis and aims to intensify human-to-human, non-human-to-human and non-human-to-non-human relationships, networks and exchanges.

Arbor at Potenziale 3 Exhibition, Innsbruck, Austria.

bl.O.serie is a living interior panelling system, a photosynthetic boiserie, that captures CO2, purifies and oxygenates the air.

1.02.04

bl.O.serie

bl.O.serie, part of the PhotoSynthetica bio-digital innovation research line, consists of porous wall-mounted 3D-printed photo-bioreactors. Each reactor is a cluster of close packed cells hosting several thousands of living Chlorella spp. cells. The system morphology and spatial composition is reminiscent of the life cycle of a unique collective green algae species, Volvox.

Volvox is unique in manifesting cellular aggregation of thousands of units in colonies that comprise up to three generations of cells. The material resolution and morphologic differentiation found in Volvox has been studied and algorithmically reproduced to design bl.O.serie.

Their collective logic materialises in the aggregation patterns of bI.O.serie components and individual cells, each hosting living cultures of microalgae. The size of the microalgae cells is comparable with the resolution of the 3D-printed reactor's cells, which have been designed and fabricated to allow sufficient airflow across the living active medium. This favours an efficient exchange thus maximising the system's ability to absorb air pollution and carbon dioxide.

bI.O.serie can be installed in any interior and on any vertical, oblique or curved wall. The living cultures will store and re-metabolise the toxins captured from the air transforming them into biomass, thus activating novel circular economies of matter, information and energy and contributing to a healthy interior environment.

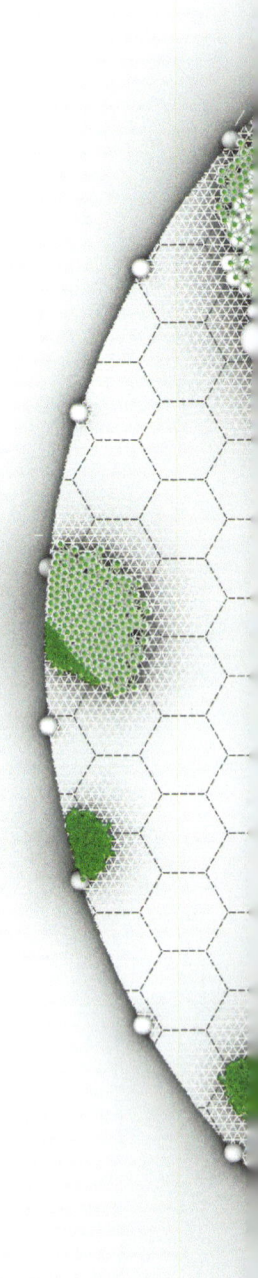

Cover page: Close-up view of the breathing wall-mounted bI.O.serie at the inauguration of the Back to Future exhibition at Museum für Kommunikation Frankfurt.

This page: Simulated aggregation of bI.O.serie cells following a close packing algorithm. The drawing shows a cellular aggregation of a virtual colony of 10,000 cells.

... the new meta-language of architecture. Patterns recognition, real-time responsiveness and artificial life.

1.02.05

meta-Folly
FRAC Orléans
permanent collection

"Upon crossing a summer field you may be surrounded by the unmistakable sound of a swarm of crickets; it is a very spatial experience as the sound appears to be moving with you. Crickets do indeed respond to the physical presence of an intruder by suddenly stopping their characteristic rubbing of wings, which produces the well-known sound. It is, we shall say, a very material kind of sound as it is the product of a physical pattern of movement articulated in space and time. Not only that but the tonality of the sound is directly influenced by the age and size of the animal as well as by its location in relationship to the listener and the consistency of the grasses and flowers that may constitute the field's green carpet. This complex set of dynamic relationships in space and time ends up producing a pattern of material sounds that we instantly recognize as characteristic of crickets." [From the original project description for the ArchiLab9 Catalogue.]

ecoLogicStudio's meta-Folly, commissioned by the FRAC Centre in Orléans and now part of its world-renowned permanent collection, is a sonic pavilion which aims to embody in architecture the spatio-temporal vibrations that we perceive as living nature. This is achieved by means of a playful perceptual game with the listener, triggering the development of a new meta-language of architecture, based on material experience, pattern recognition and real-time responsiveness.

meta-Folly demonstrates that this can be achieved with the sole use of cheap ready-made materials reassembled and activated by means of digital computation. The project thus explores the other frontier of the bio-digital paradigm, where the biologic component does not literally inhabit architecture but is materialised in the interaction between human and artificial components.

In its essence, the meta-Folly can be understood as a field of digitally materialised sensitivity which agitates, via electrical stimuli, a proliferation of 300 piezo-triggered analogue buzzers that are modulated in four different tones. Programmed to operate at variable frequencies, the buzzers react to the speed of visitors' movements around the Folly, developing ripples of sound that bounce back and forth until dissolution, synchronisation or complete interference. The convolutedness of the geometry thus leads to the emergence of unique sonic niches to be decoded by human ears within the Folly.

Cover page: meta-Folly engaging in a playful interaction with the visitors of the CCCB in Barcelona.

This page: meta-Folly at ArchiLab9, FRAC Centre, Orléans. Its physical reach is extended by the sensitivity of its six hubs, equipped with proximity sensors and able to capture the public's presence at a distance of up to five meters.

The meta-Folly has an organic meta-spherical shape with three interactive access points and two main environments; a central standing niche and two smaller side niches. These three types of access guarantee three different types of experiences of the space allowing access to three visitors at a time.

Computational cyber-artificiality of meta-Folly challenges our assumptions about what constitutes the boundary between the natural and the artificial realms. As an architectural prototype the Folly is completely synthetic, yet we perceive it as akin to a complex living system composed of multiple interacting living organisms. In this sense, the project may be said to offer a vision of Arcadia as embodied in a completely abstract, mathematical and synthetic model.

The project thus draws a path toward the future convergence of cybernetics and environmental psychology, digital computational design and craftsmanship, and radical ecosystemic thinking. It is a prototype for architecture in the Anthropocene. The outcome may appear to be an improbable assemblage of urban paraphernalia (recycled polypropylene, hacked sound kits, steel rods, chameleon nano-flakes, and so on), but within it we find a new aesthetic and spatial milieu, a new form of material life.

The architectural precedent for this project is found in the tradition of the folly within the playful English landscape gardens of the 18th century, where architecture was a device for establishing a new relationship with the natural environment. Among the many kinds of folly, the 'grotto' is perhaps the most intriguing. Typically accessed by boat across a man-made lake, the grotto was an immersive environment that faked the spatial effects of a natural cave and attempted to reproduce the exoticism of the Mediterranean. Light and sound reflections played a crucial role in enhancing the intimacy of the environment for visitors and stimulating more intense interactions. Of course the metaFolly is not designed to evoke distant memories of a romantic trip in the Mediterranean, rather it attempts to discover in the grey artificiality of the urban and in the abstraction of the computational, a new naturalness.

There is no escaping the fact that 'trash' in its broadest sense is an essential condition of our present urban society, one that far exceeds the conventional notion of waste: urban trash incorporates a multi-layered assemblage of products, landscapes, media content, attitudes and lifestyles. We can re-engineer trash, we can recycle trash or we can prevent trash from being produced, but with this project we attempted to accept 'trashiness' as a condition of our times, of our society, of our way of producing and consuming, of designing and of manufacturing and building. We decided to take 'trash' seriously, with mathematical rigour, digital precision, craft and care. This ambition drove the prototyping of the meta-Folly. Its manufacture was a meticulous process of manipulating multiple forms of cheap, mass-produced, recycled and in some cases leftover material.

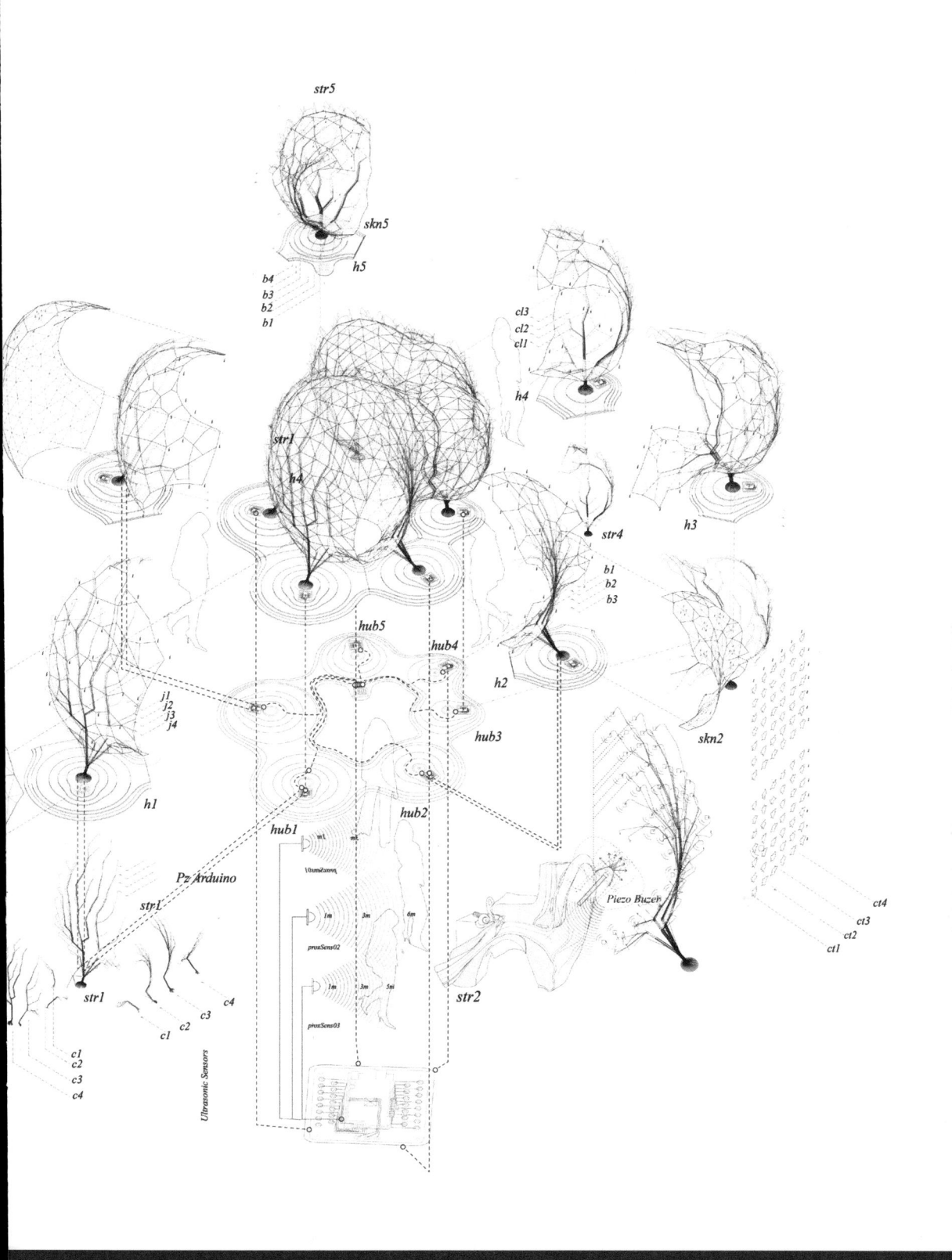

Exploded axonometric diagram describing the key hardware components of the six hubs of meta-Folly. Each of the six hubs is subdivided into interactive clusters made of five, six or seven tiles. Each cluster is equipped with one active tile, containing within itself an active tendril.

The meta-Folly's skin is developed as a mesh of approximately 1,300 tiles. Each individual tile is developed through a custom-designed script which also measures the length of each fibre in the supporting structure. A bespoke wire-bending machine was manufactured to bend the wires individually in three dimensions according to the angles and lengths provided by the script. Laser-cut connectors hold the fibres in place at each bifurcation. The bundles start at the base of each hub terminating in each of the vertices of the tiles.

In terms of fabrication and material technologies, such an approach leads to a process of 'slow-prototyping' and the development of dedicated know-how for gathering off-the-shelf and recycled items and detailing their assemblage. Through the definition of a system of transformations of found objects and manufactured connections between recycled parts we finally pieced the pavilion together. This slow-prototyping protocol was deployed in a month long workshop where a team of young architects became digital craftsmen. This rewarding albeit laborious process cemented our belief that today there is no need for 'fast' architecture, just as there is no need for fast food. Rather we shall try to practise a form of slow architecture that operates like a swarm, or in the swarm, and is capable of converting a multitude of simple instructions into an emergent meta-language of forms, movements and effects.

A fibrous system wraps and supports meta-Folly. The system, which is composed of steel rods, was developed algorithmically as a collection of bundled minimal paths. The paths are the materialisation of the trajectories of both structural loads and information travelling from each tile to the ground, all the way to the base of each of the six hubs. Minimal paths have been computed from each hub to all of the tiles, and the denser bundles are materialised by means of bundled steel rod. There are 700 rods in the whole pavilion.

Spatial proximity and the pace of movement of visitors is sensed by meta-Folly and translated by its microprocessors into time delays that activate the sound loop of each one of the 300 piezo buzzers. This simple mechanism is activated in real time by multiple individuals creating an emergent complex behaviour of sound as they move within the Folly. Each individual sound is of the simplest nature, however their interaction in time and space can produce an unlimited variety of sound inflections. At the opening show during ArchiLab9 at the FRAC in Orléans, the behavioural response of the buzzers was set to mimic a swarm of crickets in a field: when no interaction was present the speakers would loop sounds in a random sequence, whereas a human presence would increase the looping time in proportion to distance from the speakers, such that the closer speakers would turn quiet for a longer time. Speed of movement would then determine the delay before the sound resumed its 'normal' looping time.

A chameleonic nano-flake pigment covers the surface of each tile. meta-Folly's skin can thus change colour from green to blue to yellow when hit by light at different angles.

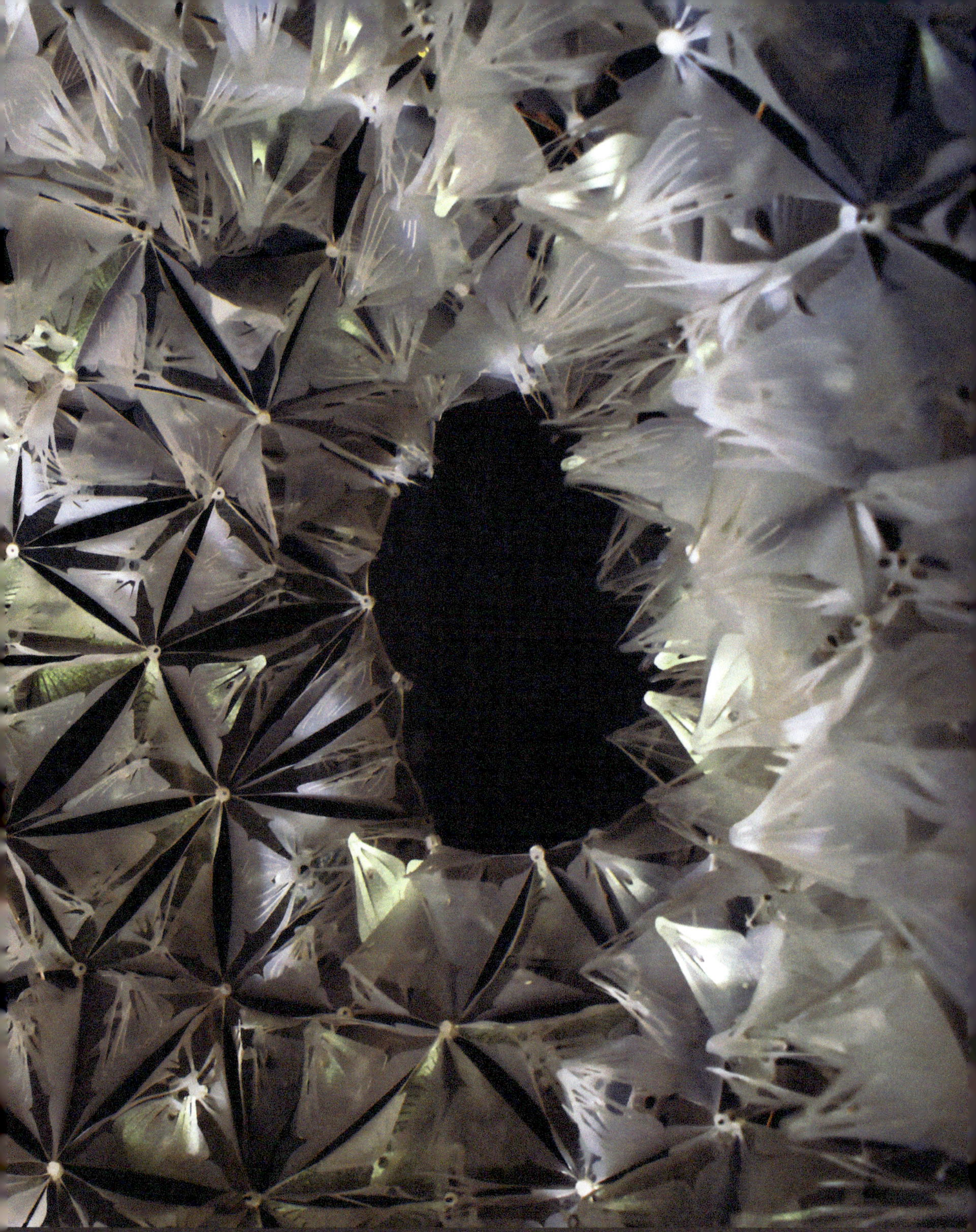

Each tendril hosts a piezo buzzer connected to a microprocessor contained in the base of each hub; the wires connecting the tendrils to the hub follow the same route as the structural fibres. Each hub's base also hosts a proximity sensor; this can sense the visitor's presence and movement and sends a signal to the processor that activates the buzzer. The visitor then hears the response and acts accordingly; his/her reaction is registered and fed back.

Each active tile contains an active tendril. The tendril is a unit of interaction and integrates a piezo electric buzzer; each tendril is connected to a microprocessor Arduino host at the base of the hub. This in turn is fed real-time data by three proximity sensors positioned at the base of each of the six hubs. There are approx. 300 active tendrils in the metaFolly that are constantly activated by the six microprocessors

Each cluster is equipped with one active tile. The tile is composed of a single laser-cut sheet, folded and held in place by two connectors supporting the internal structure of the tendril. The system is compact, minimises material use and protects the electronic parts from external agents. The tendril hosts the buzzer in a self-contained unit that terminates with a simple plug.

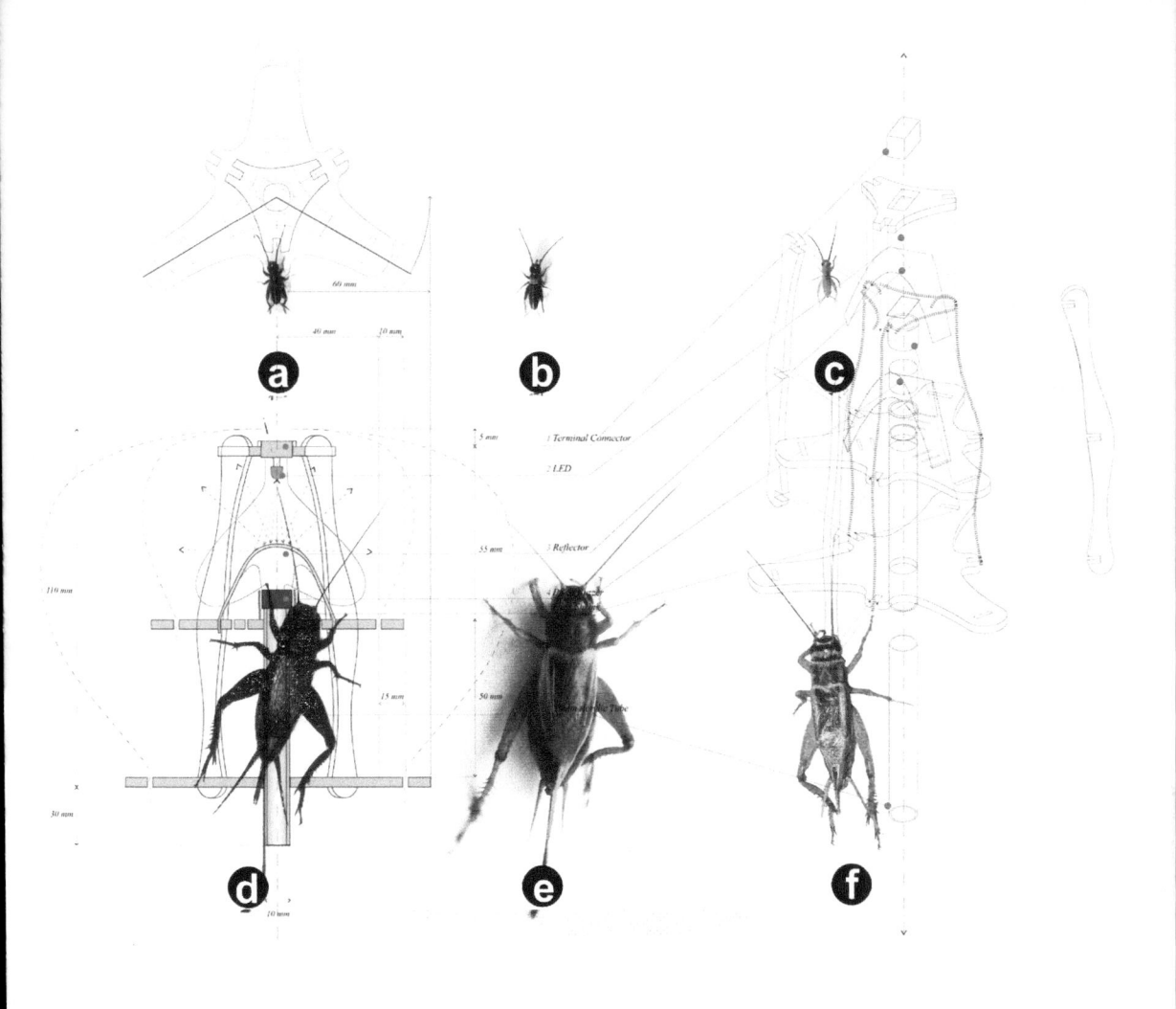

The diagram illustrates the concept of material or analogic sound that inspires the design of the active tendrils of the metaFolly. The musical tone of each tendril derives from the lengths of the acrylic tube at its core. This is inspired by the sound of living crickets whose tone also depends on the size and frequency of flapping of their wings.

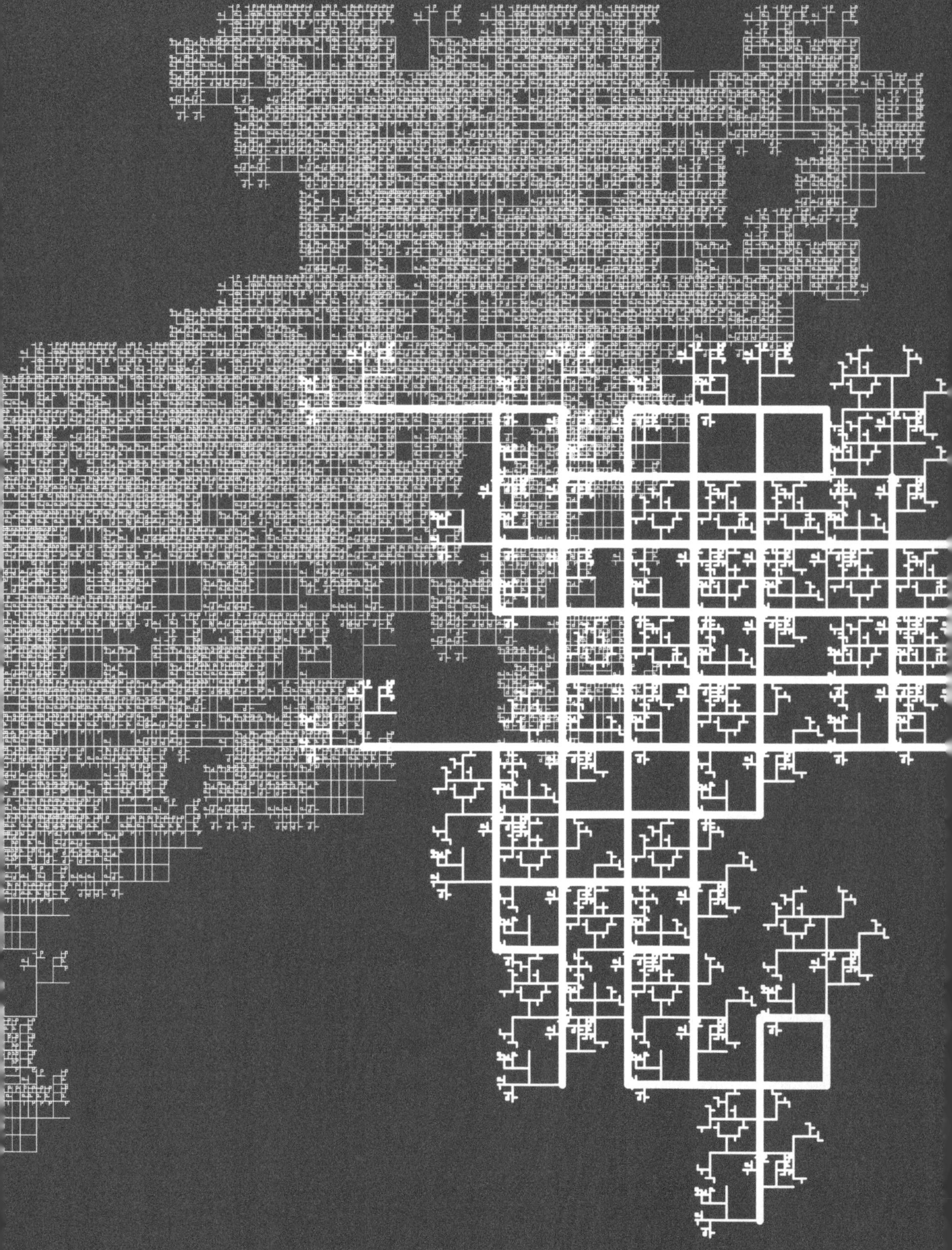

ENVIRONMENTALISM BEYOND IDEOLOGY: REPROGRAMMING THE BLUE-GREEN CITY

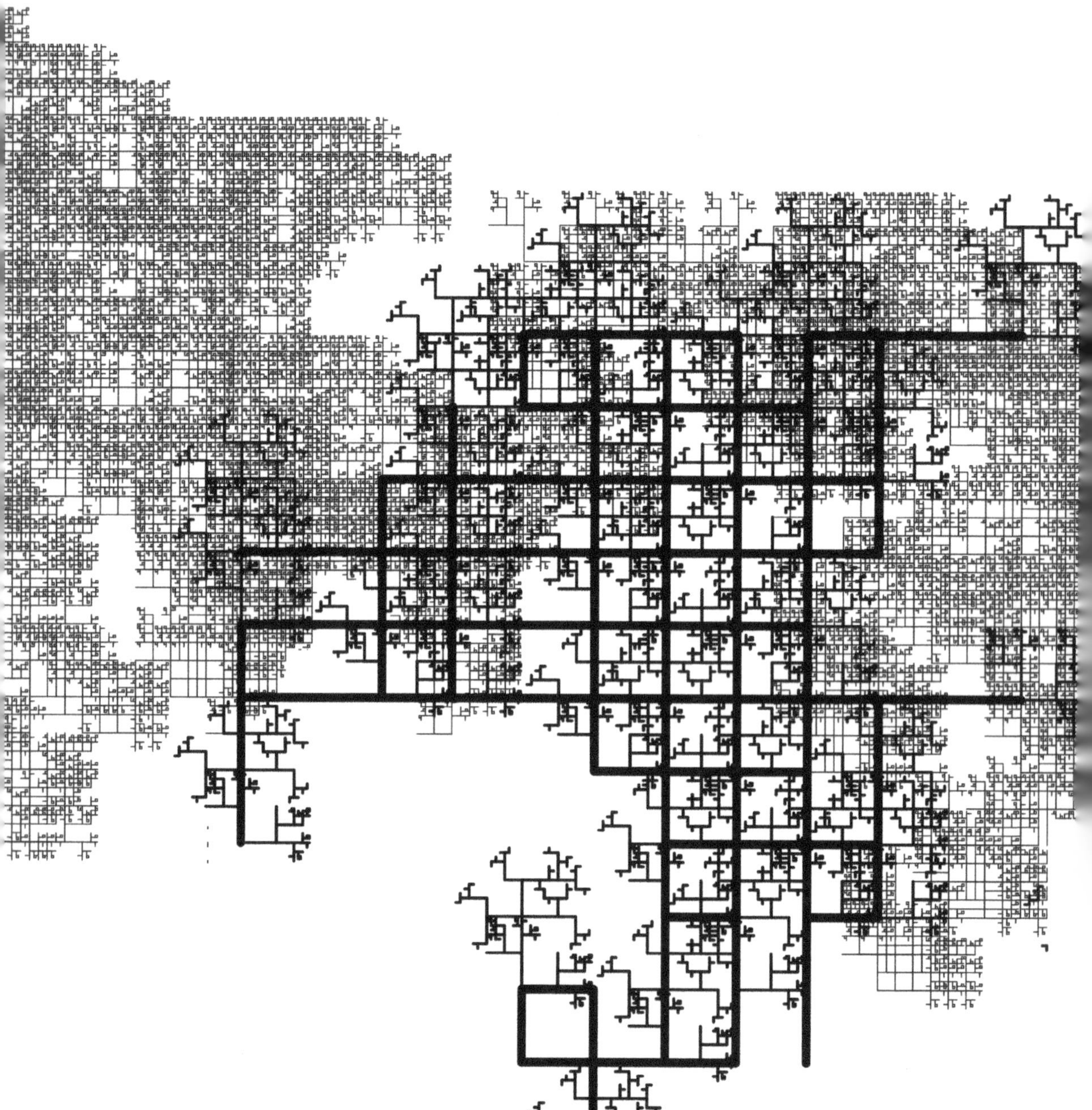

❞ The Urbansphere is our real contemporary biosphere, not its antithesis. It is our living habitat. A turbulent system, far from equilibrium, dotted with alternating moments of crisis, decay, erasure and origination. The Urbansphere is endowed with capabilities for profound reorganisation, transformation and novelty. Despite its artificiality, it embodies all the dynamic traits of the natural biosphere.

Contemporary environmentalism often seems unable to escape its ideological pitfalls and recognise the inextricable complexity of our techno-human condition, along with the implications of that complexity for our understanding of the urban environment. Cities, humans' biggest and most extraordinary invention, remain, from this perspective, antagonistic toward nature. As a consequence, the re-greening of cities is consistently presented as necessary if we are to rebalance their impact on the biosphere.

This attitude has led to a conservative cultural tendency that seeks the ruralisation of the urban landscape, a tendency that is underpinned by a flawed reading of both the history of cities and of the science that describes their impact on the natural environment. To illustrate this point, the authors will use Milan as a case study. As we will see, Milan clearly demonstrates how a flawed perspective inevitably leads to conservative positions when it comes to formulating new models of urbanisation. This is especially apparent in the conventional failure to substantially redefine notions of scale, building typology, urban zoning and the evolution of urban technological infrastructures.

After introducing the case of Milan, this chapter investigates an alternative conceptual reading of contemporary urbanity based upon the realisation that Anthropocene urbanisation has global reach that affects every living ecosystem on the planet, whether directly or indirectly. From this perspective it is possible to formulate a more abstract model of the city that makes no ideological distinction between man-made technologies and nature's biological systems; it is the 'Urbansphere', our contemporary augmented version of the Biosphere.

Milan is Italy's second largest agricultural municipality.[1] Its peri-urban territory is still dotted with countless farms, creating a web that extends all around the city's southern edge to the Expo site, an area now named the Parco Sud. This peculiar characteristic has favoured a strand of rhetoric that tends to portray the agricultural area of Milan as the new ecological frontier of the city, with features varying from the picturesque (in the most peripheral farms often seen as agro-tourist or agro-cultural retreats) to the agro-chic (in the more urban locations, typically transformed into fashionable organic eateries).

While progressive in appearance, both tendencies are conservative in character and respond to the global rhetoric of contemporary environmentalism, which leaves them unable to escape the ideological conception of nature as a fundamentally balanced and benign system currently perturbed and unsettled by urban proliferation, associated as it is with industrialisation.

This narrative ignores the fact that nature is not a fundamentally balanced system. Our current techno-human condition, largely the product of industrialisation and technological development, has been fuelled by the discovery of an enormous reservoir of energy. This energy has been stored in the form of crude oil and other fossil fuels, themselves the product of one of the largest catastrophes in the history of the Earth, which occurred before humans appeared. The biomass resulting from mass extinctions and the widespread destruction and death of organisms was trapped under many strata of rock for millions of years until our civilisation accessed it. Ignoring the complexity of such geological cycles is problematic as it results in a reductionist and, ironically, machine-like version of nature that is sanitised of all those aspects that do not fit the benign green portrayal.

In a recent article titled "How to change the course of human history (at least, the part that's already happened)" authors David Graeber and David Wengrow make a pointed remark about the consequences of the ideological misreading of history and of how it affects our ability to change the course of our civilisation:

> "for centuries, we have been telling ourselves a simple story about the origins of social inequality. For most of their history, humans lived in tiny egalitarian bands of

---

1 In Italy, Milan is the municipality with the largest agricultural surface area after Rome, according to ISTAT data from 2012.

See the Open Data website of the Comune di Milano. The metropolitan area of Milan has a highly developed agricultural territory: more than half of its surface area is agricultural land or forestry land. Parco Sud is Europe's largest agricultural park and contains more than one thousand operational agricultural businesses.

See Cattivelli, V. (2014). L'esperienza degli orti urbani nel comune di Milano: Una lettura attraverso gli open data comunali. *Agriregionieuropa* 10, art. 39.

hunter-gatherers. Then came farming, which brought with it private property, and then the rise of cities which meant the emergence of civilization properly speaking. Most see civilization, hence inequality, as a tragic necessity. Some dream of returning to a past utopia, of finding an industrial equivalent to 'primitive communism', or even, in extreme cases, of destroying everything, and going back to being foragers again. But no one challenges the basic structure of the story. There is a fundamental problem with this narrative. It isn't true. […] Our species did not, in fact, spend most of its history in tiny bands; agriculture did not mark an irreversible threshold in social evolution; the first cities were often robustly egalitarian."[2]

[2] Graeber, D., & Wengrow, D. (2018, 2 March). How to change the course of human history. *Eurozine*.

Similarly, industrialisation is often portrayed as the breaking point, the moment when society became disconnected from nature and when nature's human-made distress began. The logical consequence of this view, as exemplified by the case of Milan, is an idealisation of the idyllic Arcadian character of agriculturalism, where the pastoral scene of the urban farm or the allotment is elevated to a symbol of a possible return to life in balance with nature, in contrast to the disconnected urban lifestyles, a product of industrialisation. So we measure our cities' interaction with the biosphere in terms of its urban footprint, ecological impact and so on.

That conventional approach is a way of framing environmental problems that is best suited to technocrats, enabling them to tinker with the numbers without properly addressing the real factors that people actually care about. As with the case of social science, mainstream environmental science is now mobilised primarily to reinforce a sense of hopelessness whereby ecological catastrophe appears an inevitable consequence of the urban condition.

However, Graeber and Wengrow have examined the anthropological research closely, and found overwhelming evidence that human societies since prehistory have always experimented with multiple alternative social structures and that the linear transition from hunter-gatherer to pastoralism to agriculture and then to industry is all but an illusion:

"In the summer months, Inuit dispersed into small patriarchal bands in pursuit of freshwater fish, caribou, and reindeer, each under the authority of a single male elder. Property was possessively marked and patriarchs exercised coercive, sometimes even tyrannical power over their kin. But in the long winter months, when seals and walrus flocked to the Arctic shore, another social structure entirely took over as Inuit gathered together to build great meeting houses of wood, whale-rib, and stone. Within them, the virtues of equality, altruism, and collective life prevailed;

[3] Graeber, D., & Wengrow, D. (2018, 2 March). How to change the course of human history. *Eurozine*.

wealth was shared; husbands and wives exchanged partners under the aegis of Sedna, the Goddess of the Seals."[3]

Similarly, today, radical differences in the urban morphogenesis of global cities are appearing, where local conditions of climate, landscape morphology as well as global flows of information, matter and energy all play a role.

In this respect, Milan's case is pertinent for two reasons. First, in the panorama of the Italian Renaissance, Milan reached its peak under the Sforza family who, under pressure to expand their territory, enacted an agricultural revolution. The high demands of feeding a growing population fuelled a technological acceleration that involved the likes of Leonardo da Vinci to build a large network of canals, named Navigli. It also promoted the construction of new farmhouses and the landscaping of highly productive terrains, still known as *marcite* or meadows. This pre-industrial yet technologically sophisticated landscape dramatically increased the productivity of the land, which became in a metabolic sense 'urban' or, as we would say today, part of the urban ecological footprint of Milan. The challenge to make Milan self-sufficient transformed its urban landscape into a productive machine, a large apparatus mediating the increasing metabolism of the city and the local landscape's natural ability to support such demands. If this was an agricultural Arcadia, it was surely a techno-human one.

This observation highlights the second point derived from Milan's case, which is the scientific and technological fallacy of ideological environmentalism. Its implications are visible in the case of the urban farms now maniacally restored as relics of a rural past. Even more so in the case of the *marcite*, which are now preserved as areas of leisure rather than converted into vertical hydroponic urban farms. This attitude not only fails to recognise the ambitions of those who built the original infrastructures, landscapes, and architectures of Milan's rural past, but also fails to actualise their relevance as part of the contemporary urban metabolism of the city and its metropolitan area.

As a consequence, no effort is made to quantitatively assess and evaluate the landscape's capacity to feed contemporary Milan or to speculate as to what kind of transformative force would be required to achieve that goal. Despite the best intentions of EXPO2015's original masterplan, none of the recent interventions in the Parco Sud, Darsena (the city's harbour), Navigli, or Expo Milano 2015 have demonstrated a clear attempt to engage with the problem of the urban footprint from a radical perspective. And by radical here we mean endowing the urban population of Milan, human and non-, with renewed transformational powers over Milan's future self-sufficiency and carbon neutrality.

Milan's case demonstrates how the eco-ideological drive can hinder the societal transformations that are necessary to evolve a more sophisticated paradigm of ecosystemic urbanisation. It has instead only invigorated other pernicious tendencies such as the architectural obsession with conservation, now unfortunately understood as a superficial instrument for branding Italy's supposedly glorious past as its best asset for the future.

Meanwhile an even more paradoxical and problematic inversion is occurring: nature, in its protected status, is reduced to a mechanised version of its real self, sanitised and flattened to a thin ornamental layer to be applied ad hoc to embellish the man-made urban landscape. Nature is reduced to a manicured green lawn, as it is in most architectural renderings, or packaged in architectural metaphors like vertical forests, urban orchards or green corridors. It is always ready to be passively consumed at any time and in any season in a perennial state of comfortable and reassuring homeostasis. This is the image of nature preferred by the corporate propaganda that also funds popular documentaries and movies of ecological Armageddon which ultimately are, in many respects, responsible for the eco-ideological drive described above.

What we are witnessing is, in other words, the establishment of a vicious cycle. Hopelessness is the predominant feeling that is purposefully instilled in the human population to maintain the status quo.

ecoLogicStudio's experiences in the public realm and as blue-green planners have reinforced the authors' understanding that our current techno-human society cannot exist outside an embedded technological paradigm that is best expressed in the complexity and richness of contemporary cities. The optimal position to engage in this complexity is found outside nature, and by breaking away from the idealised condition of illusory ecological balance that the concept of nature has come to symbolise. In doing so we can recognise the urban condition beyond the physical boundaries of the city, and in all its forms and contradictions.

Territories like landfills, wastewater treatment plants, water desalination plants, energy farms, as well as agri-urban terrains, acquire new significance because they are the terrains in which our contemporary urban condition finds its most radical expression, they become urbanity's crucial nodes. This is where the concept of Urbansphere becomes instrumental, defined as the global apparatus of contemporary urbanity, a stack of informational, material and energetic networks that spans the globe, with no beginning and no end. Enmeshed as they are with one another, these networks sustain all forms of urban life, artificial, biotic and a-biotic. The Urbansphere is our real contemporary biosphere, not its antithesis.

It is our living habitat. A turbulent system, far from equilibrium, dotted with alternating moments of crisis, decay, erasure and origination. The Urbansphere is endowed with capabilities for profound reorganisation, transformation and novelty. Despite its artificiality, it embodies all the dynamic traits of the natural Biosphere.

It is therefore not at all surprising that the Urbansphere, while being man-made and engineered primarily to support humankind, does appear to be surprisingly efficient in hosting and fostering all kinds of non-human life. Large cities have evolved as the new habitat of an increasingly diverse population of animals, insects and other form of life that were expelled from farmlands and other agri-territories.

A study by botanist Wolfram Kunick showed that countless animal species are found in all corners of our urban environments and have adapted to coexist with man-made artificial urban landscapes. The botanist searched the parks, allotments and backyards of Berlin to discover that within the city more species of both flora and fauna are to be found than in the agricultural areas outside the inner city boundaries. Kunick counted 424 different plant species in the urban area of Berlin alone. In contrast, only 357 lived in intensively used agricultural areas in Berlin's surroundings. A similar surprise was brought to light by the count of animal species. There are 130 breeding bird species in Berlin alone, including some that, like Zilpzalp or Golden Oriole, have become rare in the wild. A similar biodiversity also exists in many other European cities.

The Urbansphere, as defined here, is indeed our contemporary augmented Biosphere.

This conception shifts the discourse about future urbanisation away from the ideological clash of city versus nature. Not only that, it provides another notable opportunity. As large amounts of Earth monitoring data become available, the study of the effects of urbanisation on the living world can be interpreted and deployed to make more informed urban planning decisions. Once liberated from ideological and disciplinary constraints, informed design decisions can affect the implementation of new policies for the profound restructure of the processes actually affecting people's urban behaviours.

In the Urbansphere, which like the Biosphere is an autopoietic system in a perpetual state of self-making, all living systems are understood as co-evolutionary. Technology is embedded into new material assemblages whose proliferation depends on their ability to foster life. Their performance is not evaluated against a single target or geared towards the resolution of a given problem but in their ability to develop patterns of life, manifested morphologically once they actualise within a specific design scenario. The aesthetic appreciation of such morphological patterns is elevated to fundamental evaluation criteria. It is the meta-language of urbanspheric design.

The description of the Urbansphere therefore relies on a language of abstract simulations not dissimilar to those already adopted by scientists to describe the Biosphere. Let us consider weather forecasting and the prediction of the formations of hurricanes: data are plotted as diagrams and maps that embody or manifest the predictive qualities of sensing. When we work at such levels of complexity and at such scales, the possibility of direct perception becomes very limited and experience becomes insufficient if not aided by predictive simulations. Visualisation in the form of maps is thus a crucial part of the process. Therefore, when data are used to make predictions and influence policy, but no visualisation of their variation in space and time is apparent, we should not doubt that we are operating in the realm of ideology. Visual maps are not immune to ideological manipulation, but they certainly provide a platform for moving beyond it.

Within the disciplines of architecture and urban design, the implementation of this scientific method in redesigning the urban realm has been pioneered by geographer and urban planner Professor Mike Batty, the founder of CASA, UCL's Centre for Advanced Spatial Analysis. Professor Batty makes extensive use of algorithms, and more recently of 'big data', to simulate city systems and visually describe their emerging qualities in space and time, both simulated and real time.

In an inspiring contribution, 'The Liquid City', to the authors' first book *Systemic Architecture*, Professor Batty writes:

> "Cities are fractal in their form and function […] they self-organise from the bottom up […]. This is design the way nature intended, and as cities enable more and more of their populations to indulge in positive decision-making, they are becoming more and more organic. Traditional planning and design that fights against such self-organisation will fail."[4]

Hence it is necessary that we aid our judgement with the use of algorithmic models, but in the age of ubiquitous computing algorithmic models have become embedded, and citizens can interact with them in real time. Professor Batty adds:

> "cities manifest a new liquidity of action, a confluence of light and speed, which we term the 'liquid city': a place where physical desires, face to face contacts, and digital deliberations provide a new nexus of innovation. Flows, networks, connections rather than inert buildings dominate

[4] Pasquero, C., & Poletto, M. (2012). *Systemic Architecture: Operating Manual for the Self-Organizing City*. London: Routledge, p.18.

this physicality as infrastructure comes to represent this liquidity which is built on layer upon layer of flux and flow."[5]

The actualisation of this 'soft' and 'wet' urban paradigm to novel architectures requires a few conceptual adjustments; among them is the necessity to index the city as an open and fluid urban terrain. That implies the abstract redefinition in computational terms of known architectural or urban typologies, dispensing with their extensive definition and their spatially rigid boundaries.

The reframing of types in ecoLogicStudio's work goes hand in hand with the redefinition of a common term, 'prototype'.[6] Prototype commonly refers to a test model for a new product; a prototype often materialises a more or less accurate representation of the final mass-produced object. We intend prototype here as a conceptual space of design and material and spatial organisation that exists beyond its idealisation as a type. In ecoLogicStudio's design practice the definition of prototype prefigures a large number of simulated architectures, or material systems, that may or may not actualise, but that are indexical of a territory and its specific qualities. This conceptual substitution of type with prototype is instrumental to this vision of 'digital liquidity' or 'wetware' as we described it earlier.

In fact, the indexical quality of the prototype allows us to probe the urban environment and re-describe it by means of simulated plans that we call 'operational fields'. Such field descriptions, substitutive of the so-called 'masterplan', are by definition, abstract and mathematical or algorithmic. Operational fields are a means to test sensitivities and to simulate urban plans capable of indexing multiple and nonhomogeneous parameters. Operational field maps actualise latent opportunities that cannot be registered by typical urban masterplans which are too biased towards geometric and figurative depiction. Their abstract or diagrammatic nature makes operational field descriptions sensitive to all kinds of urban stimuli—biologic, morphologic and social.

Professor Batty refers to contemporary cities as "more organic". This claim has a profound meaning that goes beyond the obvious metaphor. It touches upon a deeper cognitive layer, where individual and collective behaviours are formed. And indeed, if we look at collective intelligence at work in biological organisms like ant colonies for instance, we can observe how complex tasks such as nest building are achieved through mechanisms of communication that utilise the environment itself as their milieu (stigmergic communication).[7]

Ants leave pheromone traces in the environment that they are engaged in shaping or constructing. The architecture of the

---

5 Pasquero, C., & Poletto, M. (2012). *Systemic Architecture: Operating Manual for the Self-Organizing City.* London: Routledge, p.18.

6 Pasquero, C., & Poletto, M. (2012). *Systemic Architecture: Operating Manual for the Self-Organizing City.* London: Routledge, p.114.

7 Tschinkel, W.R. (2004). The nest architecture of the Florida harvester ant, *Pogonomyrmex badius. Journal of Insect Science,* 4, art. 21.

nest is, therefore, internalised in their communication system and its construction is by necessity a self-regulating process: it is automatically responsive to the changes the building actions are promoting in the surrounding environment. Therefore, a top-down masterplan is no longer necessary—and as a matter of fact does not exist in ant colonies—substituted as it is by a continuous form of adaptation via local communication mediated by the environment. The colony and its milieu are co-evolutionary. Even more significant is how extensive qualities such as the formation of walls or size of rooms are emergent from the intensive fields of pheromone and through the function of spatial and material constraints such as the ants' body size, granularity of soil and so on.

ecoLogicStudio's design projects presented in this chapter embrace this paradigm shift in thinking and make a concrete effort to dispense with the logic of 'zoning' that still dominates most urban green plans.

Within the discipline of architecture and urban design the pioneering work of architect Frei Otto is worth further examination here.[8] Otto studied for several years the processes of occupation and self-organisation of large territories as they occur in various realms, from a-biotic systems to the animal kingdom. He compared those results with the behaviour of various civilisations and how systems of connectivity would evolve within them. To simulate such processes at a time of limited computational power he built specifically designed apparatus for computing from soap bubbles, sand, ink droplets and other analogue media. Such 'material computation' operates morphologically and in relationship to a specific substratum or medium. Otto was able to compute using completely analogue technologies the emergence of path systems, in particular he defined a special category of path systems, the so-called 'minimising detour' networks. These networks are unique. They produce connections among points that are optimal in terms of the energy expenditure required to connect them. The notion of energy here implies the existence of a substratum and of an 'organism' moving over it. Otto discovered that the patterns emerging from his experiments exhibited self-similarity in non-human as well as in human systems.

Frei Otto's findings are relevant now more than ever, especially in light of the need for designing systemic models of co-evolution in the Urbansphere.

In particular, ecoLogicStudio has recently been involved in the so called 'blue-green' planning of the cities of Aarhus, Denmark and Tallinn, Estonia, where the principles described above have found practical application.

8 Otto, F. (2008). *Occupying and Connecting.* Stuttgart: Axel Menges.

9 Leonardsen, L. (2017). The blue and green city: The water story of Copenhagen. http://civiccommons.us/wp-content/uploads/2017/08/Copenhagen-BlueGreen-City-keynote-Leonardsen.pdf

The Scandinavian planning model is among the most developed in building resilience into urban planning and takes the living substratum of the city as its foundation. For example, the city of Copenhagen has recently adopted a blue-green plan for the next 20 years[9] and that plan is focused on providing a more efficient wet infrastructure, operating at two very distinct scales: at the larger scale its urban and peri-urban green systems are designed to enhance green-ways and park connectivity, while its many small-scale installations implement the strategy in a typically urban realm, for instance by adjusting street and park design in order to retain rainwater and by rediscovering its canals and water gardens. However, the plan still adopts a typological language defined by rigid spatial metrics and clearly segregated scalar and programmatic dimensions. No recognition of what Professor Batty describes as the fractal nature of cities emerges here, despite the many waterways, the tidal zone and the lagoons that together are among the most fractal systems in the Biosphere.

To overcome these shortcomings, ecoLogicStudio has developed an urban design workflow (which will be described in more detail in later chapters) that involves deploying satellite imagery to scan the urban terrain at a resolution of approximately 10x10 metres, thus enabling accurate data of soil porosity, vegetation density and urban topography to be collected. Such measurements can then be plotted onto intensive urban field maps that are not dissimilar in nature from barometric weather maps. The existence of a variation in their scalar field implies a series of relationships to that variation. For example: given a field of air pressures there is always a direction at any point in which the pressure difference is maximal, and that direction will be the prevailing wind direction (vector) at that location; or, given the topographic field of a hill, there is always a direction at any point in which this tendency is maximal. The intensity of this maximum together with its direction at every point describes a vector field. If droplets of rain fall on the hill, then this vector field will define their average flow pattern. Using computational algorithms our projects can simulate these patterns for millions and even billions of points across the urban terrain and can thus describe the emerging flow pattern as it evolves with every variation of the parameters in the model. The local territorial processes that define the composition and porosity of the urban 'crust' at any given point in the urban landscape are the primary focus in the design process and can thus be understood as co-evolutionary to large-scale urban infrastructures.

Algorithmic simulations of this kind also produce emerging organisational patterns at intermediate scales that blur typical scalar domains and define new fractal orders. It thus becomes possible to take design decisions and test policies at any one of these scalar levels and simulate their impact at all other levels simultaneously.

A similar example can be made regarding green networks, for which the benchmark is the city of Malmo.[10] Malmo's plan is based on a green taxonomy that defines spatial hierarchies within the city's green network. This network is connected via 16 green corridors that each pass through a wide variety of natural spaces, from lagoons to wetlands, large and small. The green areas are catalogued in terms of scale, but are still defined extensively by area.

In the proposal for the city of Tallinn, ecoLogicStudio's workflow drops all reference to such extensive definitions. Satellite vision is deployed to measure vegetation density on a computational grid dividing Tallinn into 10x10 metre pixels. Minimised path systems are computed, connecting each building in Tallinn to the closest pixel containing biomass above a certain scalar threshold. The paths in each system are computed along the existing road network and the resulting path systems are then differentiated and widened accordingly to the level of biological traffic running through each link.
The resulting plan creates a green capillary network serving every building in the city with the city's constantly evolving biological potential.

In this proposal there is no 'top-down' structure, symmetry or scalar segregation. What emerges instead is a fractal and distributed network of green links carrying all kinds of informational, material and energetic signals. The green plan is itself as alive as the systems it is trying to describe and manage, forming a large-scale urban wetware.

In conclusion, if conventional urban greening ideology does little to advance our conception of a sustainable and carbon neutral city, architecture must embrace urbanity in all of its contradictory aspects to engender significant change. A radical vision for a future green city should be as removed as possible from any picturesque or pseudo-rural depiction. It needs an abstract, computational dimension to simulate its material, informational and energetic processes in high resolution and in real time, thus creating a planning medium that provides freedom to engender mechanisms of transformation and thereby effect a new urban morphogenesis.

10 O'Byrne, D. (2018). Malmö, Sweden: The City of Parks. http://depts.washington.edu/open2100/Resources/1_OpenSpaceSystems/Open_Space_Systems/Malmo_Case_Study.pdf

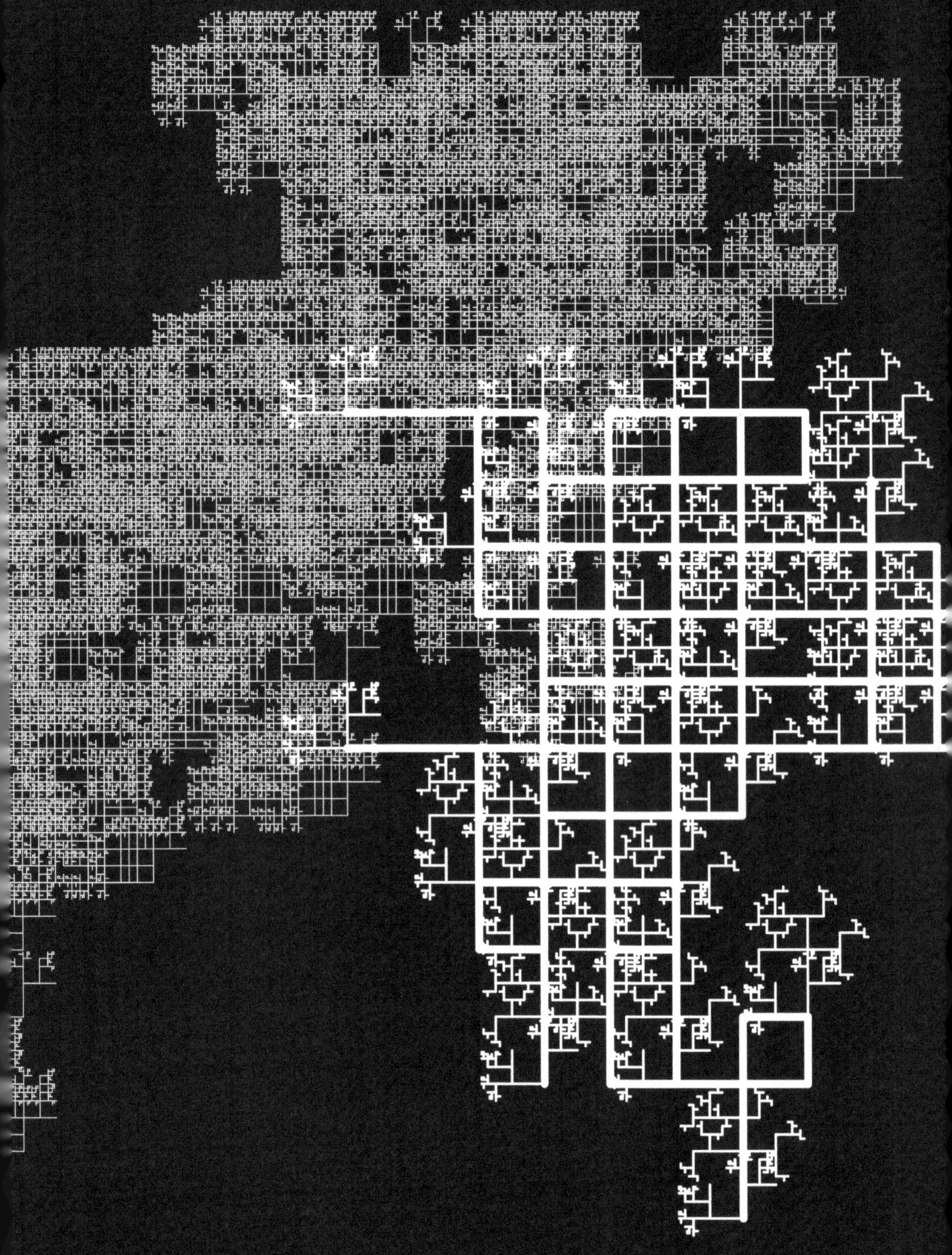

Projects 1.03

CYBER-GARDENING THE CITY

1.03.01 Urban Algae Canopy, the Metropolitan proto-Garden 194

1.03.02 Urban Algae Folly, the Aarhus Wet City plan 202

1.03.03 AirBubble COP26 218

🎵 The Urban Algae Canopy is a bio-digital living pavilion converting solar energy into biomass and oxygen.

1.03.01

Urban Algae Canopy and the Milan Metropolitan proto-Garden

ecoLogicStudio first developed a set of operational fields for Milan in the project 'Metropolitan proto-Garden'. The project comprised a video installation commissioned by Luca Molinari that engaged with discussions around the masterplan for Expo 2015 that had been proposed by the advisory team of Stefano Boeri, Richard Burdett, Jacques Herzog, Joan Busquets and William McDonough.

ecoLogicStudio employed the concept of a diffuse Expo and proposed a new infrastructure of advanced urban gardening stations, all equipped with digital sensing and actuating mechanisms and fitted with synthetically engineered microalgae colonies; collectively such prototypical assemblages formed a Metropolitan proto-Garden capable of reactivating and re-metabolising Milan's network of farms and its historically productive landscapes.

The Metropolitan proto-Garden challenges the separation between the urban and the rural condition as well as the separation between urban networks and urban landscape. The existing urban morphology of Milan is augmented with real-time simulations of air pollution, vegetation density, wetness and access via public transportation, depicting new fields of opportunity for breeding a distributed network of biotechnological gardens. Presented as an open-source interface, the Metropolitan proto-Garden simulates, visualises and operates this new biotechnological stratum in real time.

For example, the relationship between air pollution and the uncontrolled proliferation of microalgae colonies that currently grow in the waterways of the city would be managed and exploited to promote urban enhanced photosynthesis, urban oxygenation and biomass production for both food and biofuel. Such bio-digital mechanisms can feed on what we now consider urban waste or pollution, thereby activating new pollution-to-food feedback loops, significantly affecting the overall urban metabolism, and reducing its footprint. Social groups and other active stakeholders can register for one of the many virtual plots that algorithmically subdivide the territory of Milan and thereby actively develop interest in specific forms of cultivation. Various descriptors (enabled by ubiquitous data-logging) provide an account of the material processes that are channelled through each plot and influence directly the urban cultivation process.

As the plots evolve into proto-Gardens, new urban systems emerge from a process of actualisation of latent social, economic and ecologic potentials. Cultivation evolves into a process of culturalisation. These potentials are recognised through the framework of the virtual plots and the computation and plotting of the related operational fields; their actualisation in a multitude of forms, spatial conditions and social communities is catalysed by the introduction of new prototypical urban and architectural devices. These devices are characterised by a specific set of performances (evapotranspiration, photosynthesis, carbon sequestration, etc.) engineered in the form of specific material organisations or systems (branching, folding, weaving). The systems are embedded with remote sensing and actuating potential (temperature, pH, radiation, proximity, etc.) and interfaced with a simulation engine (digital interface).

The project implements an infrastructure that aims to enable the proliferation of new urban practices of what we call 'bio-digital cultivation', that is intensive urban agriculture augmented by digital and biotechnologies. Such proliferation would contribute to a dramatic increase in the city's ability to grow food, produce energy, recycle waste and metabolise pollutants.

Cover page: Close-up view of the photosynthetic Urban Algae Canopy shading the main public space of the Future Food District at the Milan EXPO2015.

This page: Metabolically, the Canopy absorbs CO2 emissions to feed the algae and oxygenates the air at a rate that is 10 times higher than a large tree can do in the same area. It also shades and cools the microclimate of the Future Food District.

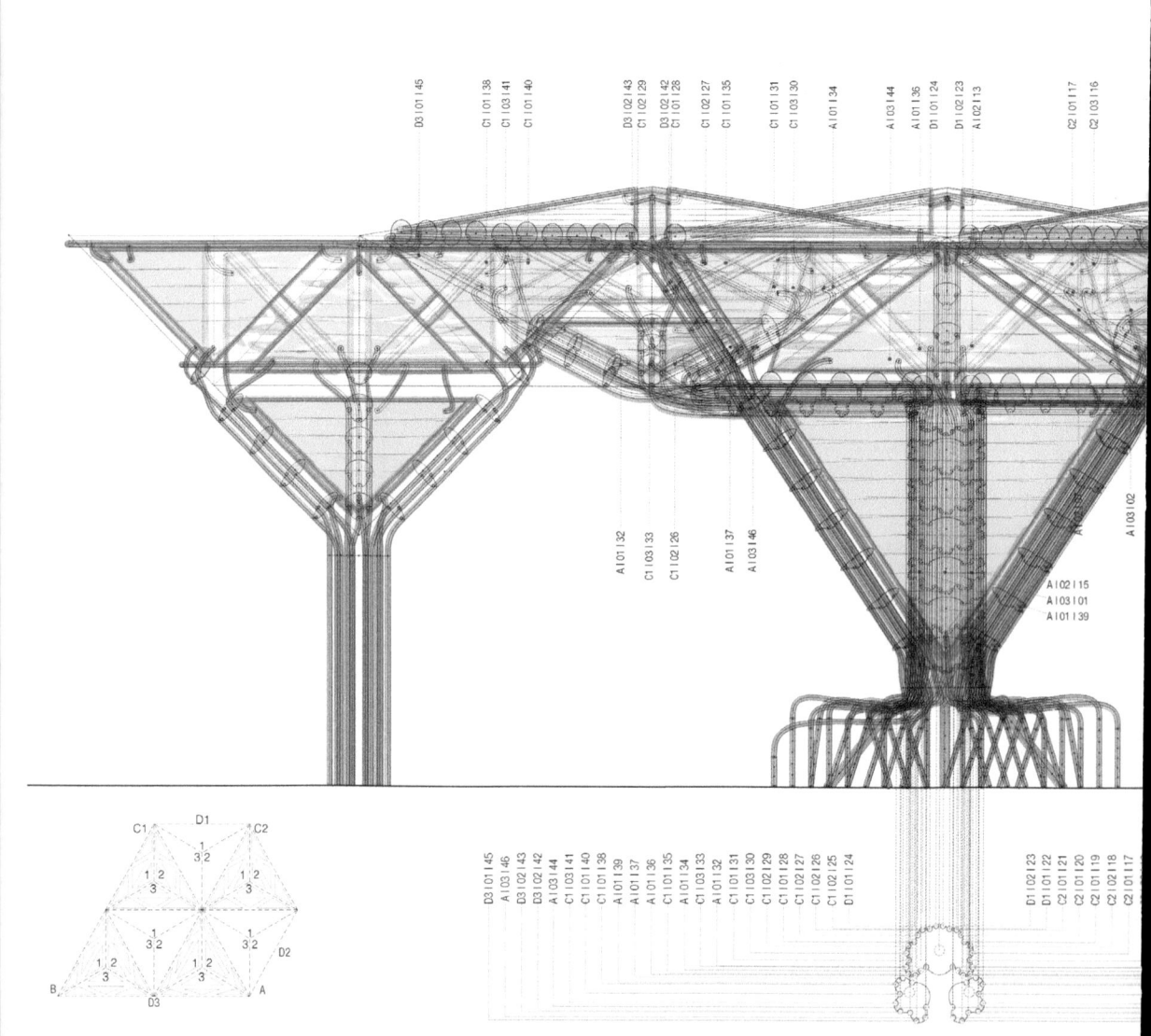

Detailed construction drawing of integrated architectural algae flow system. The innovative architecture of the Urban Algae Canopy originates from the evolution of the well known ETFE architecture membrane. In this instance it has the ability to carry the flow of liquid medium and to provide an ideal habitat for living Spirulina cultures.

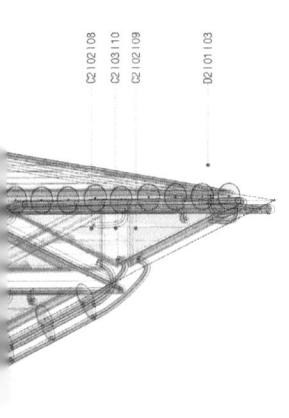

One of the experimental proto-Gardens, the 'Urban Algae Canopy', was designed, engineered and built by ecoLogicStudio as part of Milano EXPO2015 in its Future Food District. This architectural folly was designed to provide an artificial habitat for the cultivation of edible microalgae colonies of Spirulina. The Canopy articulates the cultivation process in space and time and integrates many technologically innovative solutions. It is the world's first architecture made of a soft ethylene tetrafluoroethylene (ETFE) skin that is engineered to host living cultures of cyanobacteria. It is also the first time such a system has been installed in a central and busy urban environment and is capable of sensing both microclimatic conditions and human responses to them.

The innovative architecture of the Urban Algae Canopy originates from the evolution of the well known ETFE architectural skin system. The ETFE system provides the ideal habitat to stimulate Spirulina's growth and also to guarantee visitors' comfort.

The translucency, colour, sound and productivity of the Canopy are the result of the complex interaction of climate, living cultures and humans, mediated by digital control systems. The Canopy is sensitive to human presence and movement by means of eight proximity sensors on the four columns, covering the entire area in and around the folly. As people move, the sensing system reads their presence and speed, and live data-feeds are transmitted to a central microprocessor that computes the status of the nine solenoids or electro-valves controlling the algal flow before it enters the ETFE cushions.

Algae flow in a closed loop, growing as they photosynthesise. As they become more dense the colour of the canopy darkens. When fully grown the cultures are deep green/blue in colour and are able to block 90% of the incoming light. Every 10–15 days harvesting is necessary: one third of the culture is filtered out to obtain fresh wet Spirulina and water is reintroduced with fresh culture medium to dilute the cultures and feed another cycle of growth.

In our research, we discovered a rich variety of water plants growing spontaneously in urban waterways and water bodies, including various algae species whose properties surprised us. We discovered that their ability to photosynthesise is 10 times greater than that of large trees and that they can produce vegetable proteins with far greater efficiency than any form of animal farming.

Urban Algae Canopy thus creates a new space for urban experimentation with these organisms. It moves beyond accepted types, expanding the range of technical and spatial solutions to a given problem. This is achieved by deploying technical innovation to re-describe the problematic field itself. Microbiologists and agronomists thus become active stakeholders in shaping the future of the urban realm. They can form new strategic partnerships with municipal and community groups to turn carbon sequestration into an opportunity to grow biomass and super-food.

The Urban Algae Canopy proposes a new architectural language that repositions farming and horticulture beyond its rural dimension and right at the core of future urbanism. Agriculture thus becomes a key tool for reprogramming the public realm of Milan, articulated through architecture in space and time.

Algae flow in a closed loop and keep growing as they photosynthesise. When fully grown they turn the canopy a deep blue-green colour. The Canopy is sensitive to human presence and movement. The interaction data are computed to operate nine solenoids controlling the algal flow as it enters the ETFE membranes. The behavior of the Canopy evolves in real time.

> The artistic dimension of architecture is a means to interface technical, technological and scientific innovation with society's future needs and sustainable values.

1.03.02

Urban Algae Folly and the Aarhus Wet City plan

In summer 2016, while Marco was an Adapt-r research fellow at the Aarhus School of Architecture, ecoLogicStudio engaged with a group of architects and activists working on a project called 'Dome of Visions'. The cultural legacy of projects like Synergia Ranch, Biosphere2 and other utopias was evident, not only in the design of the Dome itself, indebted as it is to Buckminster Fuller's geodesic concept, but also in the idea of testing at 1:1 scale new forms of co-evolution with nature.

The pervasive atmosphere of the Dome was that of a neo-hippy commune, devoted to small-scale cultivation of vegetables and architectural experimentation. The premises of the projects were interesting, but the apparent lack of interest in the application of more sophisticated forms of technical and scientific expertise was problematic. The project was presented as a form of resistance towards the corporate machine of urban development, but it finally become apparent that its limitations were another manifestation of conventional green ideology at work.

Cover page: Drone shot of the Urban Algae Folly in the harbor of Aarhus, European Capital of Culture 2017.

This page: Urban Algae Folly incubating living algae cultures from Aarhus harbor. The sculpture becomes a beacon for the natural biodiversity of the Aarhus aquatic microbiome.

While moving from similar cultural premises, our project titled 'Aarhus Wet City', developed in collaboration with Aarhus city council and Aarhus Capital of Culture 2017, proposes an alternative action plan. It tests the application of digital and biotechnologies to the urban realm in the city of Aarhus port to uncover the design opportunities latent in its microbiome. The project comprises an urban prototype and a digital interface.

The prototype, the 'Urban Algae Folly', operates as an incubator for the micro-ecologies present in the city's urban waterways, revealing the unique complexity of such ecosystems. It proposes a future vision where microorganisms contribute to the city's own expanded metabolism by filtering urban wastewater, digesting organic waste, and growing bio-fuels and super-food for human and animal consumption. The Urban Algae Folly, hosts a total of 400 litres of living cultures. At best, it can absorb 500g of CO_2 per day and produce 375g of O_2. Its ability to absorb CO_2 is equivalent to that of eight large trees which is the amount of trees needed to provide the annual O_2 needs of a single person.

The algorithmic urban plan for Aarhus Wet City expands on the potential of the Folly approach for the future blue-green layers of the city of Aarhus. At the urban scale, biological and aquatic species (part of what local planners call the blue-green layers of Aarhus) are networks, their proliferation determined by a large number of feedback loops that affect multiple systems including man-made ones. A future resilient blue-green masterplan must recognise these existent biological networks and enable their co-evolution with urban infrastructures.

Typically, however, urban designers and planners tend to rely on a typological segregation of functions. So, for instance, we often refer to parks, green ways, urban forests, fountains, ponds and so on as clearly defined and spatially demarcated types. But trees in a park, in reality, are a network, and so are their roots connected via biological/informational networks operated by mycelia. Such networks do not stop at the end of the park area but continue both above and below ground. Insects, worms and other animals, as well as wind and rain, contribute to the resilience of these networks by providing alternative channels of communication for seeds, pollen, and so on. Architecture too shall be used to expand these channels of communication.

A close-up view of the Urban Algae Folly photosynthetic membrane highlighting the naturally occurring bio-film formations.

Close-up view of the photosynthetic ETFE membranes and the bio-inspired laser-cut supporting structure.

Digital technologies and satellite monitoring devices enable designers to map and draw every single tree in a city and their exact geographic location—in Aarhus, at the time of mapping in 2017, there were 141,765 trees. If we then run an algorithm that draws a line connecting all the trees that sit within a certain distance from each other we can explore the nature and morphology of these urban bio-networks. We can visualise urban parks as areas with a high degree of connectivity and conversely detect zones where the network has a poor level of connectivity and new planting may be critical, providing precious missing links to increase its resilience.

Such networks can also be overlapped with topographical analysis and rainwater flow patterns to see how water and tree networks spatially and physically interact. Foliage and roots systems are among the most powerful tools available to us for controlling and preventing flash flooding, erosion and other water-related hazards. Therefore, a strong and co-evolutionary blue and green infrastructure provides a resilient substratum for our cities to grow and thrive in the future.

The virtual model of our Aarhus Wet City plan also maps the interaction between this biotic substratum and the urban networks that we, humans, have built, such as water and energy infrastructures, cycling routes and other public and private transportation networks. We have included in our virtual model of blue-green Aarhus an abstract version of the main road network of the city, where every junction is a node and every road or path a link. We can then map every single building as a dot, a starting or end node in the network of roads and paths.

Sensible planning guidelines suggest that every one of us should live in close proximity to nature and cycle or walk along natural features on our way to work or back home. We can simulate this strategy by running an algorithm to compute the shortest path from our home, workplace or restaurant to the closest tree or cluster of trees of a certain minimum density. This algorithm runs for every household in the city and creates a map of overall proximity and inter-connectivity of our man-made infrastructures and the natural biological urban networks.

Urban Algae Folly, collective urban cultivation on the opening night of Aarhus, Capital of Culture 2017.

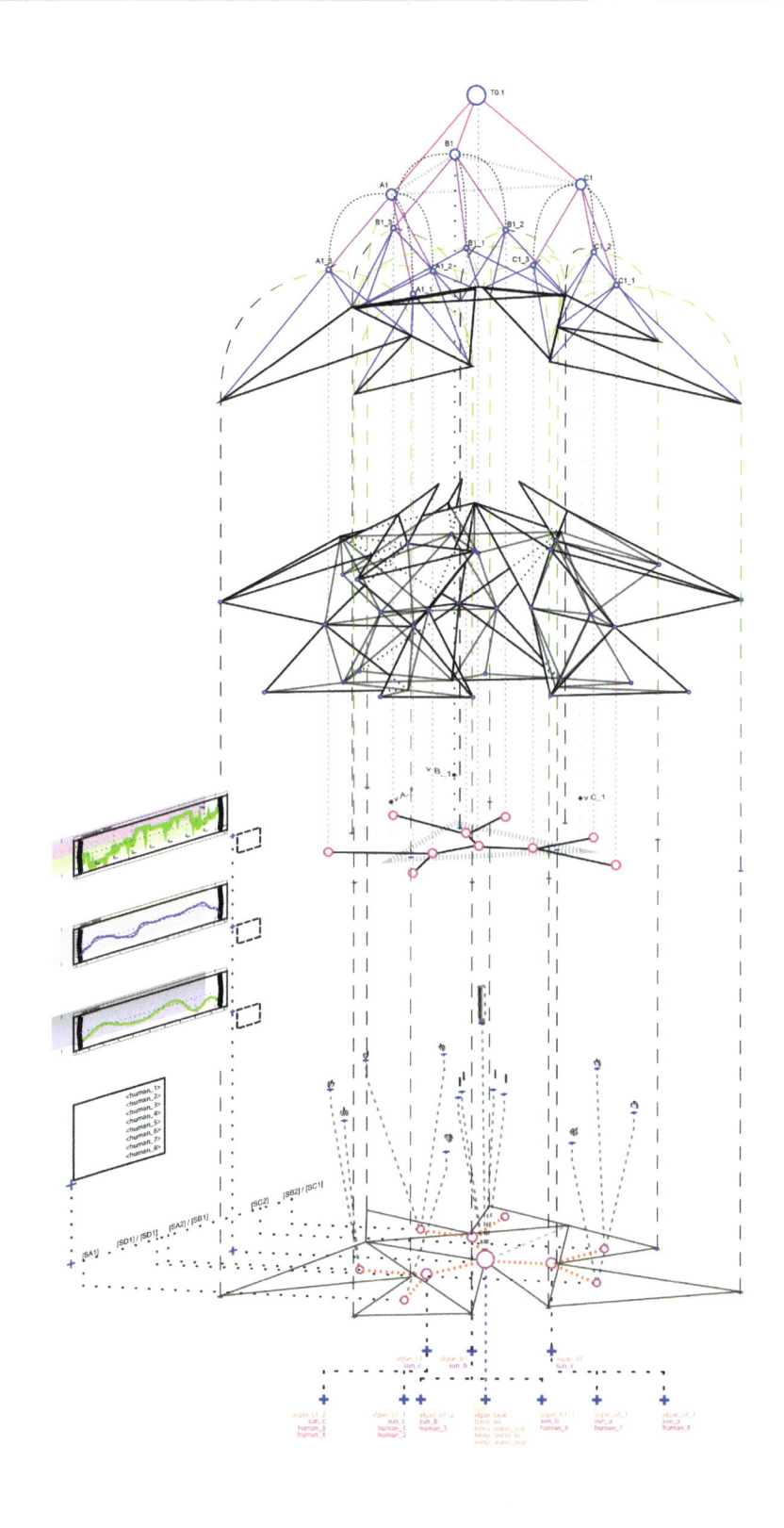

Cybernetic diagram of Urban Algae Folly describing the relationship between formal, informational and biological components. From the top: algae flow system, substructure, sensors, anchor points and coding relationship.

Since we are now in the age of ubiquitous computing we are equipped with devices connecting citizens in real time to their surrounding urban environment; our algorithm can therefore run continuously processing in real time ever-updating input data sets. Direct citizens' involvement can then be harvested not as a commentary to a proposed plan (as in public consultations) but as an active agent of planning.

Like a virtual botanical garden, we propose that every urban tree should be equipped with a digital tag that enables every citizen to check their state in real time, reading information about their location, species, age, size, management, CO_2 absorption, protein content and energy value. The digitalisation of urban trees expands the ideal of the tree into a virtual multiplicity interfering with multiple urban infrastructures. Morphological and botanical differences become relevant and are considered for their urban performance. As such single trees may be clustered while other minor biologically active species may also be included; these may include individually cultivated ones such as those in allotments and other urban gardens.

Each citizen of Aarhus could then enter his or her green-blue tag, including his or her location in the city, and generate a minimal network with connections between that location and other blue-green nodes (trees and other tagged blue-green systems) nearby. This personalised virtual garden could extend as far as each one of us may typically walk or cycle, or it could plot the equivalent blue-green systems necessary to offset the ecological footprint of our personal life. The emerging operational fields are thus participatory, distributed and co-evolutionary blue-green urban plans that redescribe the city's morphology from a new perspective; that is, from its capacity to self-sustain by means of its living land and aquatic systems.

When we ran the first algorithmic simulations on test sites in Aarhus we quickly learned just how inadequate the current biotic infrastructure is, especially in the new high-density developments of Aarhus's harbour. We also visualised, spatially, morphologically and quantitatively, the impact that biotech innovation could have at that scale. These simulations alter the overall patterns of interaction and stimulate new behaviours that feature in the evolving blue-green plan.

Currently the informatisation of urban blue-green systems is one of the fastest emerging sectors in the smart city trend. However, this research project suggests that societal benefits are best achieved when design and planning strategies take advantage of digital design innovation in order to provide a participatory framework. Smartness without participation leads to fragile and control-obsessive regimes, which in turn favour the emergence of ideological approaches such as eco-conservativism; and that is why we consider this project to be a test bed for participatory frameworks in bio-digital design rather than simply a quest for the application of new green-tech devices.

The ambition of the Urban Algae Folly is therefore to use the artistic dimension of architecture as a means to interface technical, technological and scientific research with society's future needs and sustainable values. For this reason, we are engaging with urban public spaces as open-air laboratories. Historically, public piazzas have been the core arenas of democracy, of public engagement and participation; places where the core values of our society were shaped and are still being shaped today. In this project the participation of the public is vital for evolving the methods and protocols of what we call 'collective urban cultivation'. The main purpose of such methods is to promote art and architecture as tools to develop resilient and sustainable future urban plans and a tangible vision of what a carbon neutral society may look like.

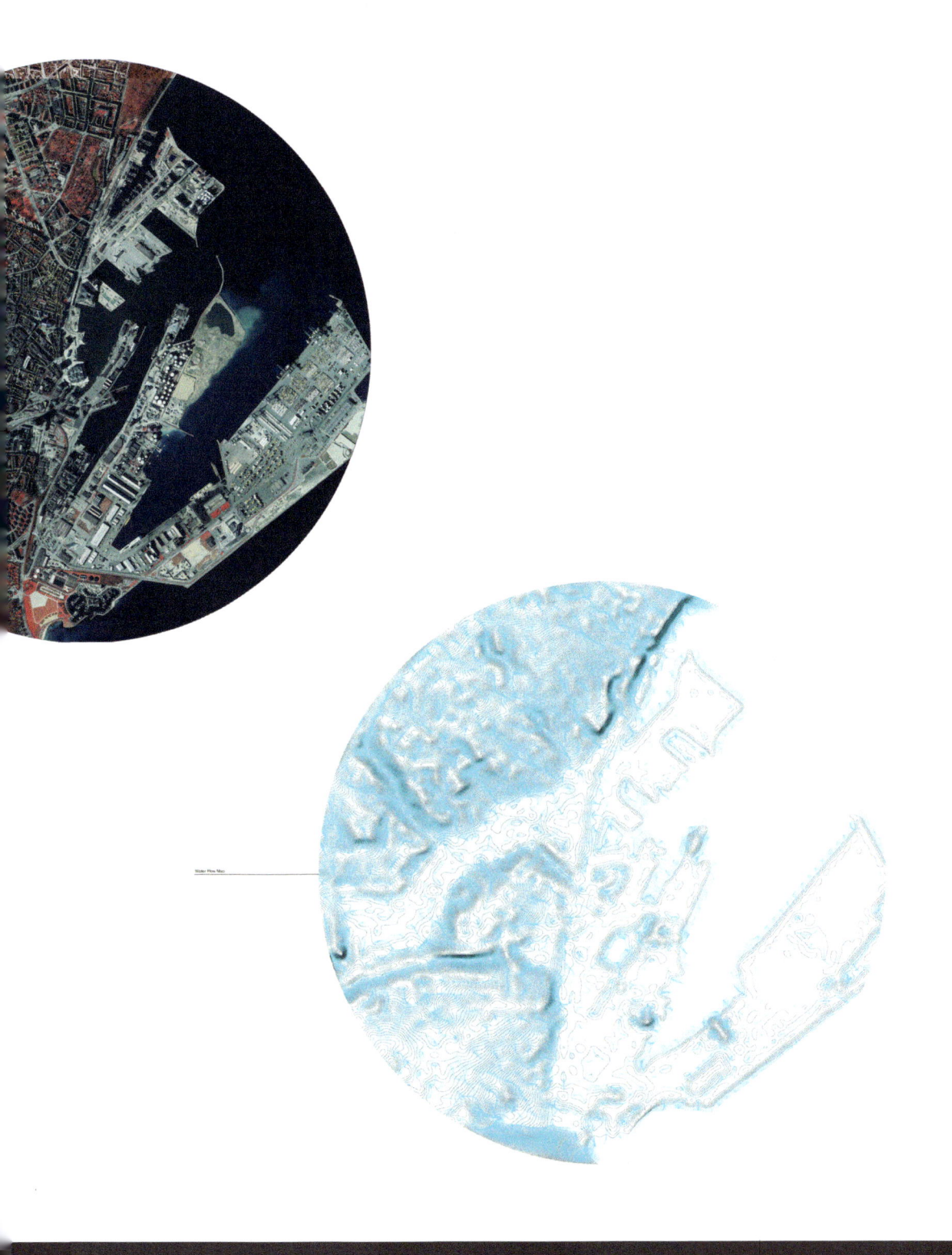

Aarhus Wet City—urban blue-green plan.

Top: The infrared satellite analysis derived from high-resolution ESA datasets, highlights information about urban wetness and photosynthetic potential.

Bottom: The urban fluid dynamic study is conducted over an accurate topographical model to determine the water flow over the surface of the city. This analysis highlights areas of potential accumulation, erosion or flooding.

Aarhus Wet City—Green Urban Networks.

This plan explores the relationship between the built environment and the natural one. Minimal paths are computed from each building to all the trees in a 500m radius. Detours are computed on the actual road network of Aarhus. The line thickness highlights the relative importance of each green link.

AirBubble brings to COP26 a concrete vision of the architecture of carbon neutrality.

1.03.03

AirBubble
COP26, Glasgow

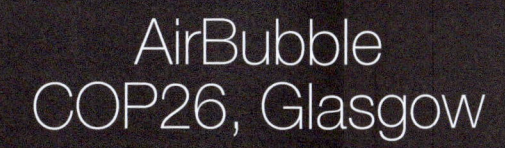

This new bio-digital project demonstrates how the advanced integration of biotechnology in the built environment can lead to a new generation of living, growing architectures. Following the successful AirBubble biotechnological playground project built in Warsaw (Poland), ecoLogicStudio developed the AirBubble air-purifying eco-machine, which was installed in front of the Glasgow Science Centre within the COP26's Green Zone area. The project has been developed in partnership with Otrivin®.

The eco-machine is made of 99% air, water and living photosynthetic air-purifying Chlorella cultures. This demonstrates how beauty and efficient ecological performance can be combined. The project encourages visitors, and especially children, to directly interact and experience the air-cleaning capabilities of microalgae cultures while immersing themselves into a bubble of freshly metabolised oxygen. The playful softness of the organic structure, akin to a gigantic bouncy jellyfish, is a direct manifestation of the biotechnology it integrates.

AirBubble air-purifying eco-machine is also ecoLogicStudio's first pneumatic bioreactor. It contains 6,000 litres of water supporting 200 litres of living Chlorella cultures that filter 100 litres of polluted urban air every minute. The air and water pressures are contained by a TPU membrane that is just 0.5mm thick, comprising only 5% in weight and 1% in volume of the overall structure. The overall strength of the structure is made possible by its three-dimensional cellular organisation. To achieve this, the fabrication process entailed the complete unfolding of the structure shape into almost 100 CNC-cut flat parts that were then welded into position to form a fully three-dimensional matrix of inflatable cells.

This design improves upon the traditional qualities of inflatable structures to create the eco-machine. The result is a responsive system, with air-purifying capability, exceptional wind resistance and unique deployability. The incredible lightness of the empty membrane makes it uniquely low in embodied carbon and minimises emissions associated with transportation, installation and dismantling.

Cover page: The AirBubble eco-machine installed in front of the Glasgow Science Centre at the entrance of COP26's Green Zone.

This page: Interior view of the AirBubble highlighting the 24 algae photo-bioreactors integrated in the pneumatic shell. The air inside AirBubble is purified of 85% of urban pollutants and filled with fresh oxygen.

The AirBubble as a bio-digital organism. Responsive lighting maps wind intensity, and human playfulness is reflected into the water ballast on the ground and the droplets of condensation on the membrane.

The outdoor membrane is monitored in real time by an array of accelerometers, sensing the wind and induced vibrations in the pneumatic structure. These sensors activate a responsive array of growth lighting that in turn support algal photosynthesis, thus increasing air purification. The entire bio-digital organism evolves a new kind of symbiosis whereby the more people play and interact within the structure, the cleaner the air becomes.

The filtering process is further enhanced by the architectural morphology of the structure. The TPU membrane—an evolution of the PhotoSynthetica urban curtain system presented in Dublin in 2018 by ecoLogicStudio—controls the microclimate inside the bubble. The inflatable membrane's doors stimulate air recirculation and natural ventilation.

AirBubble air-purifying eco-machine combines a lightweight inflatable technology with 24 photo-bioreactors (12 on each side) that are hosted in the inflatable system to create a unique microclimate inside the structure. A constant air circulation stream absorbs six core pollutants: fine particulates PM2.5 and PM10; ground level ozone (O3); nitrogen dioxide (NO2); sulphur dioxide (SO2); and carbon monoxide (CO). The project is capable of absorbing 97% of the nitrogen and 75% of the particulate matter in the air. AirBubble air-purifying eco-machine is a tangible vision of how a net zero civilisation can clean its polluted air, produce energy, grow food and construct its buildings over the next 30 years, starting now.

Interior view of AirBubble in Glasgow. AirBubble is an aquatic organism hosting 5,000 litres of water supporting 250 litres of living Chlorella cultures.

Detail of the soft TPU photo-bioreactors of AirBubble. Evaporation fills the inner gap in the membrane. Condensed droplets deposit on the inner membrane providing natural solar radiation diffusion to the living cultures and the visitors.

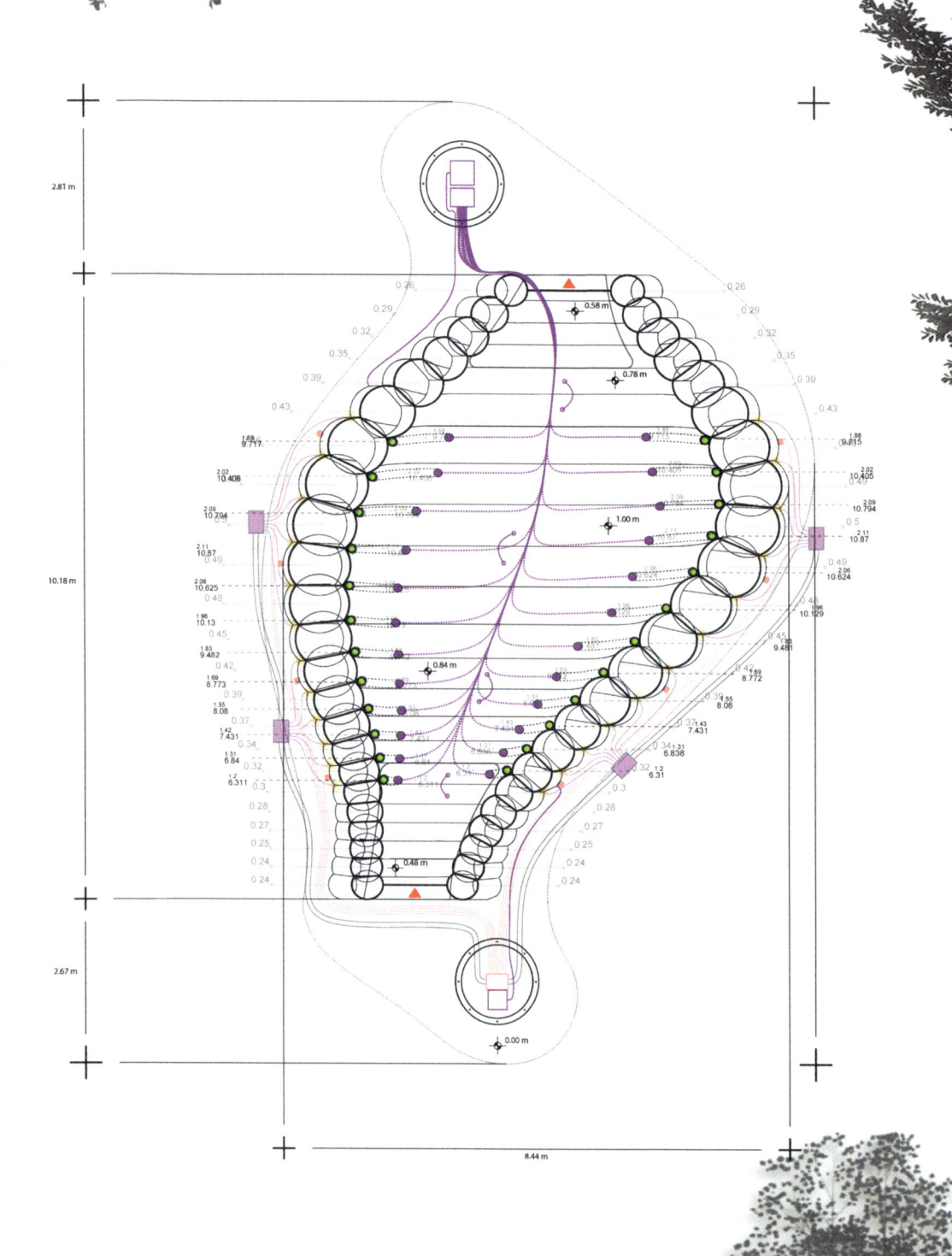

Systems diagram of AirBubble eco-machine. The diagram highlights the organic integration of the structural cellular membranes, the photo-bioreactors, the aeration system, the sensing and interactive lighting systems.

AirBubble creates its own inner microclimate of pure oxygen, diffused light and natural bubbling sound. It constantly changes appearance to offset variation in the surrounding environment and offers a soothing experience of urban delight.

PART 2

DEEP GREEN

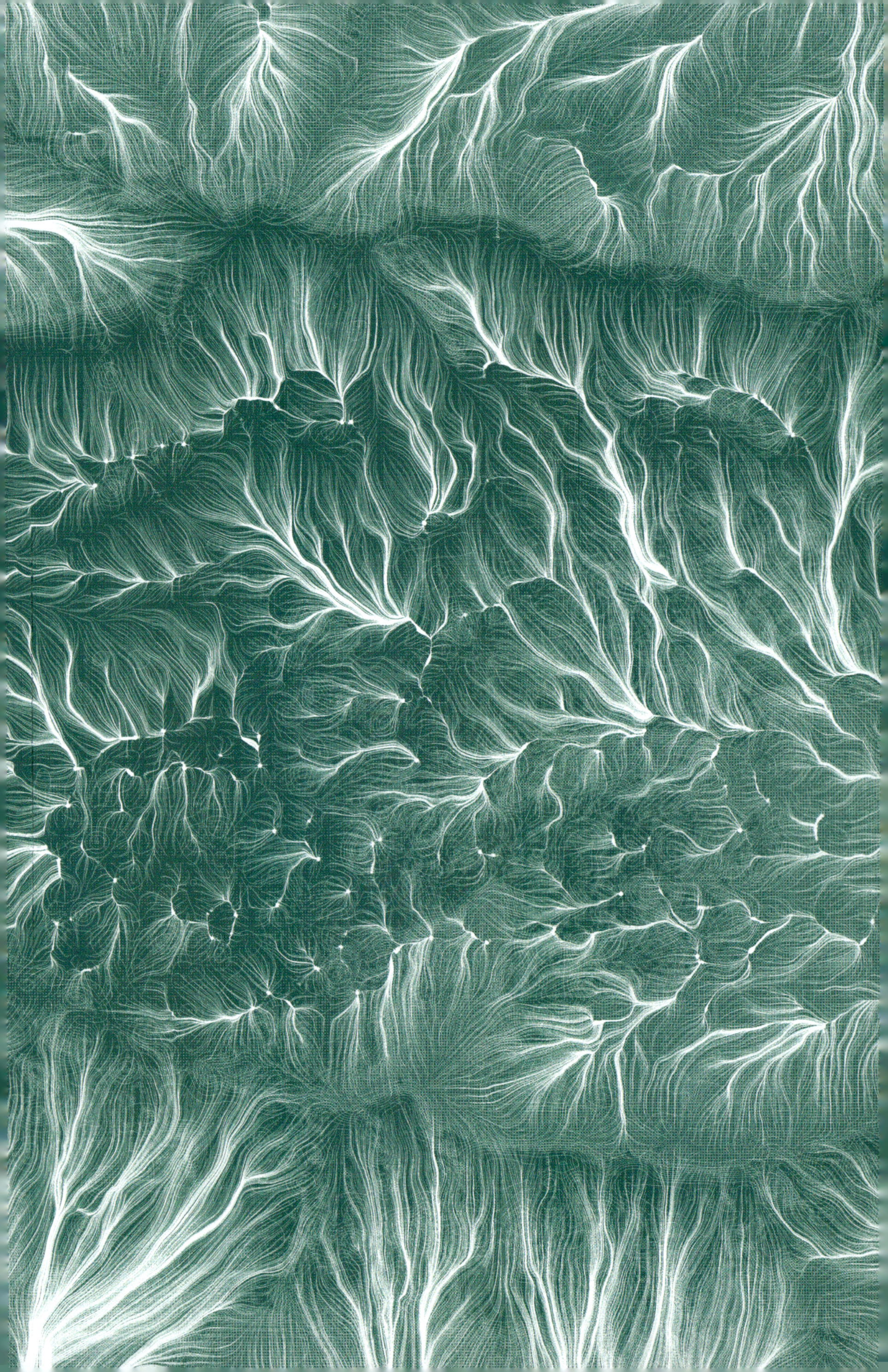

THE POLYCEPHALUM: A JOURNEY THROUGH ARCHITECTURE, BIOLOGY AND COGNITION

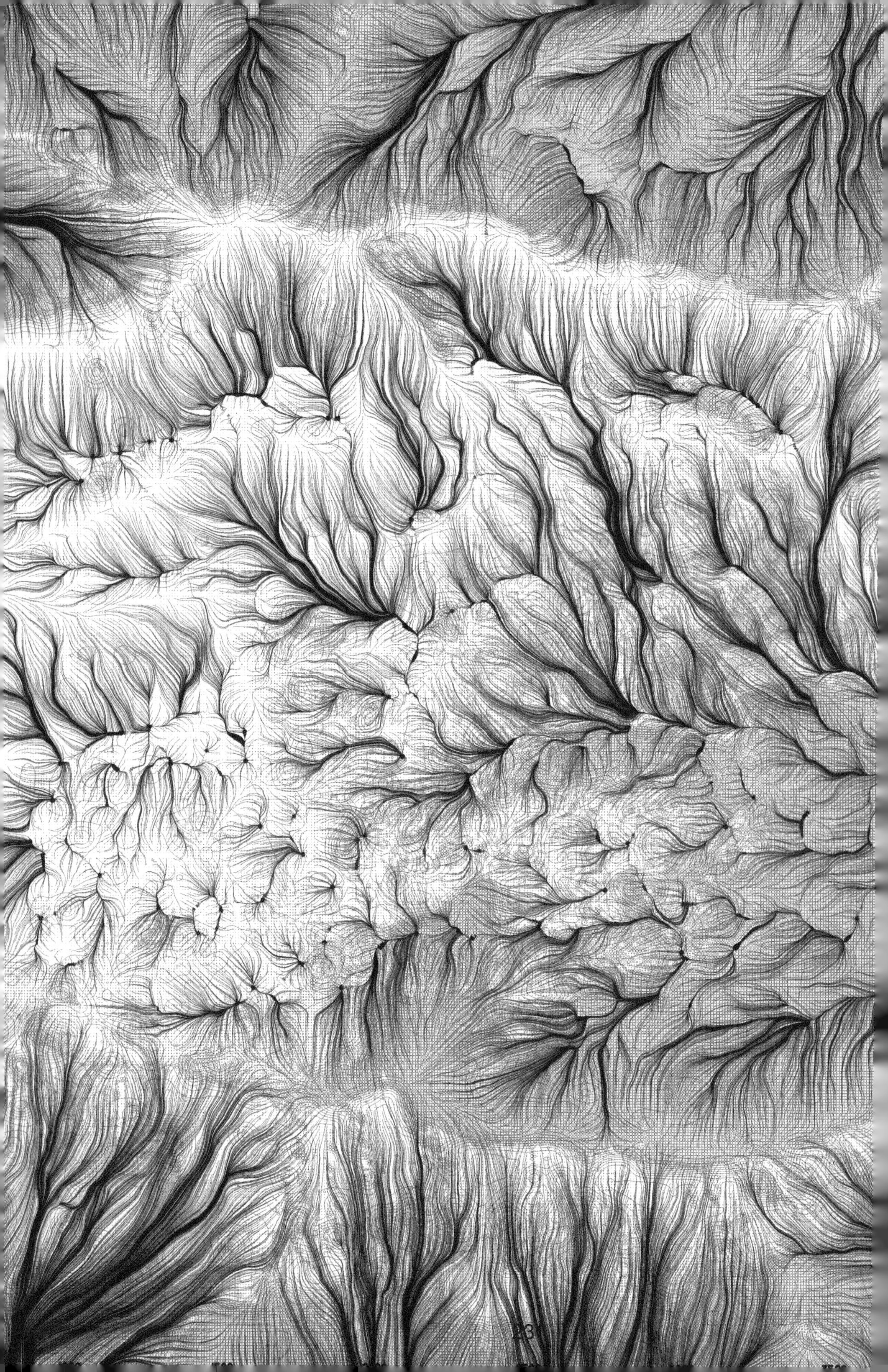
23

'Aesthetic' is intended here to mean a language of non-verbal communication, a metalanguage, which now expresses a greater ecological agency.

This second part of the book, titled 'Deep Green', presents a series of essays and design projects driven by ecoLogicStudio's broader investigation into a new aesthetic for ecological architecture.

Part 2 also articulates in more depth the design and research methodology that the authors have developed through their design practice, ecoLogicStudio, and their teaching and research work at the Bartlett School of Architecture (UCL) and at Innsbruck University. Given the multiplicity of platforms and, critically, of minds at work, this extended design innovation framework is called here a 'Polycephalum'.

The theoretical premises of the Polycephalum sit at the intersection of contemporary philosophy and post-modern science and translate findings from these disciplines into the realm of architecture and urban design, focusing on the controversial relationship between ethics and aesthetics.

This journey departs from the seminal work of anthropologist and cybernetician Gregory Bateson and his definition of ecology. It then moves to philosophers Slavoj Zizek and Timothy Morton and their visions of an aesthetic that emerges out of a renewed concept of nature.

Once a correlation between ecology and aesthetic is established the discussion will move on to exploring the design implications of this approach, looking at the meaning of the notion of material computation in the work of Frei Otto, of bio-computation in the work of Andrew Adamatzky and, finally, of embodied cognition in the work of Andy Clark.

Polycephalum expresses a multi-headed paradigm that encompasses a philosophical and an architectural design method. Polycephalum architecture, accordingly to ecoLogicStudio's body of works, interprets ecologies as cognitive systems and embodies them in a distributed design practice. As such the work of ecoLogicStudio is often seen as ambivalent: on the one hand it may appear inspired primarily by the aesthetics of nature, in a kind of formal biophilia, while on the other hand it is a technological apparatus that really 'works' or performs like a living system. The question that is often asked is whether it really attempts to perform naturally or at least as effectively as nature does.

These observations and critiques generally seem motivated by a genuine drive to understand ecoLogicStudio's practice in more depth. However, their reasoning is also anchored in modernist dualisms of form versus function and aesthetic versus ethic. Polycephalum architecture attempts to develop an aesthetic language for architecture beyond such positivistic dualisms.

Polycephalum architecture mobilises multiple forms of intelligence, both human and non-human, to redefine the architectural and urban realms. The research presented here has engaged with species such as bacteria (cyanobacteria), spiders (tarantula), silkworms (*Bombyx mori*) and slime mould (*Physarum polycephalum*), as well as with generative algorithms (GAN). All operate as actors in the design process. On this view, these organisms, together with the machines and all the other communication devices involved in the process of design become, alongside human beings, co-authors that contribute to the social dimension of design.

Polycephalum aesthetics in architecture

Dealing with a complex milieu in which multiple degrees of stability, instability and diversity coexist transforms the act of designing, increasing its potential to designate novel realms for intervention. In this era of catastrophic global climate change and perennial crisis, this kind of disciplinary expansion is necessary and long overdue.

The authors have developed their ideas in relationship to their long-standing interest in ecology and aesthetics and upon first encountering Gregory Bateson's book *Steps to an Ecology of Mind*, which experience dates back to the mid-1990s, before they engaged with architecture. According to Bateson, we live in a world populated by ecologies of mind, and our functioning as humans within these ecologies can only pass through a combination of logical and meta-logical communication channels, linking us with the world that we inhabit. Bateson wrote:

"mere purposive rationality unaided by such phenomena as art, religion, dream, and the like, is necessarily pathogenic and destructive of life."[1]

1 Bateson, G. (1972). *Steps to an Ecology of Mind: Collected Essays in Anthropology, Psychiatry, Evolution, and Epistemology.* Chicago: University of Chicago Press.

Bateson's writings were not simply an attack on traditional science, rather they underlined that reality consists of interlocking feedback loops and that through our rational understanding we are only partially able to grasp them. Consequently, reality may be pathogenically misread. It is only by combining the conscious and the unconscious grasp of complexity that we see the overall system and, eventually, imagine how to interact with it and co-evolve. According to Harries-Jones:

> "Bateson realised far ahead of his contemporaries that the primary source of error in ecological science lay in a false presumption of an ability to 'control' and 'manage ecosystems through quantitative measurement."[2]

Bateson argued that patterns are a form of analogue computation and a significant expression of cybernetic communication or meta-language. In an embryogenetic process for example, this analogue computation expresses itself in the form of fluid patterns, flowing in between the womb and the embryo in order to influence the specific growth of the embryo itself. A morphogenetic process emerges literally out of this material interaction. Socially, the aesthetic, as a form of material interaction, becomes a way of expressing, exploring and developing such meta-languages. Bateson argued:

> "So, by 'aesthetic', I mean responsiveness to the pattern which connects. The pattern which connects is a meta-pattern. It is a pattern of patterns. It is that meta-pattern which defines the vast generalisation that indeed it is the patterns that connect."[3]

The aesthetic becomes here a means to establish a cybernetic conversation within which human and non-human ecologies constitute co-evolutionary systems, a form of extended mind. Following Bateson's line of reasoning it is Slavoj Zizek and Timothy Morton writings that more recently have become particularly relevant to the authors' line of reasoning. They each define a view of ecology without nature, suggesting, albeit rather differently, a greater role for aesthetics in reframing ecological issues. Their approach articulates the shift from a problem-solving framework to a cybernetic one.

> "So while we campaign to make our world 'cleaner' and less toxic, less harmful to sentient beings, our philosophical adventure should in some way be quite the reverse. We should be finding ways to stick around with the sticky mess we are in and that we are, making thinking dirtier, identifying with ugliness [...] dark ecology."[4]

Morton thus argues that we cannot sanitise or 're-green' global ecosystems, including cities; rather, in order to fully address the issue of disrupted ecologies we should expand

[2] Harries-Jones, P. (1995). *A Recursive Vision: Ecological Understanding and Gregory Bateson.* Toronto: University of Toronto Press, p.117.

[3] Bateson, G. (1977). Box 6 Manuscripts 'Mind and Nature', unpublished, p.10

[4] Morton, T. (2007). *Ecology without Nature: Rethinking Environmental Aesthetics.* Cambridge, MA: Harvard University Press, p.93.

our consciousness to their 'darker side'. Following a similar line of thought, Zizek recognises the paradox of the current condition of society that he defines with the term 'disavowal', arguing that we know very well what the threats of an imminent ecological catastrophe are, but we still do not act energetically upon them because we cannot rationally believe them.

In other words, even if media bombard us with the horrible consequences of climate change, most of us in our daily life, have almost no experience of those consequences. That is why true ecologists should look for "trash and not for trees", which is to say that they should bring to the forefront of our collective conscious the hidden by-products of our civilisation. Rather than cover them in green propaganda, the challenge is to find their aesthetic dimension thus giving them meaning. Zizek writes:

> "There is more and more of this trash, and I think even the greatest challenge is to discover trash as an aesthetic object. [...] We should develop a much more terrifying new abstract materialism, a kind of mathematical universe where there is nothing, there are just formulas, technical forms and so on, and the difficult thing is to find poetry, spirituality, in this dimension."[5]

5 Zizek, S. (2009). Ecology. In A. Taylor (ed.), *Examined Life: Excursions with Contemporary Thinkers.* New York, New Press, pp.155–184.

During this time of global climate change and health crisis we have become more aware of the reality that today no ecosystem is unaffected by human actions. Our current stage of technological evolution, notably in the form of ubiquitous computing, synthetic biology and artificial intelligence, is opening scenarios where traditional dichotomies such as natural and artificial, material and digital, human and non-human become obsolete. In other words, we live in a paradoxical condition where indeed our impact on the Biosphere is total but our ability to comprehend and control the effects on our civilisation is diminished.

For instance, the technological apparatus of ubiquitous computation that allows us to monitor and compute the changes in the global climate also, inevitably, contribute to the intensification of those changes in unknown ways due to their intense energy consumption and carbon footprint. Thus far, an increase in computational power and distributed intelligence has generated a parallel increase in carbon footprint. This equation has remained unchanged throughout all recent 'revolutions' including cryptocurrencies, non-fungible tokens, the metaverse and artificial intelligence. Moreover, it will remain unchanged until, at last, an emerging form of non-human intelligence becomes capable of self-regulating its own impact on the systems of the Urbansphere.

This intelligence will most probably be a form of Polycephalum. It will have to be a distributed, spatial and embedded computational process. It will develop patterns of material computation and stimulate the emergence of new visual computational languages, revealing emerging territorial and urban narratives that are profoundly 'natural'. Yet, perhaps paradoxically, their naturalness will be illusionary, in the sense that we are already witnessing when struggling to define what is natural and what is artificial about our contemporary world.

From this perspective, global cities gain a new relevance. Despite being large man-made systems often depicted as the antithesis of nature, cities develop patterns of growth (or shrinkage) that recall natural formations of a radically different kind. Cities appear as complex synthetic organisms. They no longer embody the model of urbanity that we inherited from modernity where zones are clearly defined, morphologically demarcated, and programmatically determined. They are a Polycephalum.

However, planners, including ecologically minded ones, still rely today on a sanitised, zoned vision of the Urbansphere, where bacteria, pollutants and other microorganisms are considered alien and are confined to the margins of urbanity. This cultural perspective still influences the global conversation about re-greening cities and re-naturalising forests as if such processes could lead to the re-balancing of a temporarily perturbed Biosphere.

To refer back to Bateson's argument, the relationship between the human and the non-human realms contributes to the definition of novel ecologies of mind and, as a consequence, of novel morphogenetic systems. Therefore, in the Anthropocene, an epoch when our civilisation has impacted on metabolic processes at a planetary scale, we are depending, perhaps paradoxically, upon non-anthropocentric forms of reasoning and intelligence.

As Bruno Latour explained in an interview published in October 2018 in the *New York Times*:

> "The medical revolution commonly attributed to the genius of Pasteur should instead be seen as a result of an association between not just doctors, nurses and hygienists, but also worms, milk, sputum, parasites, cows and farms. Science was social then, not merely because it was performed by people [this, he thought, was a reductive misunderstanding of the word 'social']; rather science was social because it brought together a multitude of human and non-human entities and harnessed their collective power to act on and transform the world."[6]

One might ask whether on this view architecture, urbanism and design in general, as spatial and material languages,

[6] Kofman, A. (2018, 25 October). Bruno Latour, the post-truth philosopher, mounts a defence of science. *New York Times*.

could play a role in enabling renewed communication and co-evolution between human and non-human systems.

Bio-computing architecture

Recent developments in evolutionary psychology, described by Anjan Chatterjee in his book *The Aesthetic Brain*,[7] demonstrate that our sense of beauty and pleasure is part of a co-evolutionary system of our mind and the surrounding environment. Our senses of beauty and pleasure have evolved as a selection mechanism. Cultivating and enhancing them enables us to compensate and integrate our logical thinking to gain a systemic view of our planet and the dramatic changes it is currently undergoing.

7 Chatterjee, A. (2013). *The Aesthetic Brain: How We Evolved to Desire Beauty and Enjoy Art*. Oxford: Oxford University Press.

In the past century, French philosopher Henri Bergson, associated with the philosophy of 'vitalism', argued for the relevance of the continuous process of life, which he called 'élan vital'. In *Creative Evolution* (1907), long before Bateson, Bergson formulated a severe critique of the positivistic way of thinking and its static way of interpreting life. For him, one of the intellectual traps of positivism was its preference for 'freezing' systems, i.e. systems in which both the spatial and the temporal dimension are reduced to constant values. These static models are 'death' accordingly to Bergson.[8] In fact, if life is nothing but élan vital, it cannot be represented in a static model alone.

8 Bergson, H. (1998). *Creative Evolution*. Mineola, NY: Dover, p.173.

Bergson presents his argument in terms of spatial variability: the positivistic science is wrong to isolate systems while studying their behaviour, because these systems exist only in relationship to the entire universe. Similarly, in terms of temporal variability, observing a living system implies taking into consideration its behaviour in all of its duration, so that by looking only at critical points scientists are eliminating life from the processes they study. In this way, Bergson underlines a contradiction between the scientific method of understanding life and the complexity of life itself. He describes this contradiction as a dichotomy between 'intuition', which he defined as the knowledge which allows us to grasp the duration and complexity of life but cannot be expressed through logical words, and 'conceptual knowledge' which he defines as an abstract and schematic language useful in pragmatic terms but unproductive in metaphysical ones.

To resolve this, Bergson proposes the concept of 'consciousness as duration':

> "Doubtless we think with only a small part of our past, but it is with our entire past including the original bent of our

soul, that we desire, we will and act. Our past, there, as a whole, is made manifest to us in its impulse, it is felt in the form of tendency, although a small part is known in the form of idea."[9]

9 Bergson, H. (1998). *Creative Evolution*. Mineola, NY: Dover.

Inorganic reality seems to be qualitatively different from the continuous motion that characterises our consciousness, but if we look carefully, we can observe the necessity of duration also in inorganic systems. By exploiting this property of inorganic life many forms of material computer have been devised over the millennia of human society. Such a form of computation harvests the natural ability of material to self-organise. Analogue computers were material assemblages capable of multiplying the effects of material computation, thus achieving enough problem-solving performance while retaining a specific aesthetic.

Crucially, the architecture of computation is today beyond the scale of human spatial perception. It is therefore the task of computational architecture to subvert this tendency. Moved by this ambition to animate architecture, ecoLogicStudio began working on 'ecoMachines' as early as 2006:

> "ecoMachines provide a material and operational framework to deal with change and transformation, the two main defining qualities of our new understanding of urban ecology; moreover they support interaction between heterogeneous systems, such as social, infrastructural, architectural and environmental ones; they allow us to sense, register and manipulate in our daily life the unfolding processes defining our cities, our houses and our artificial environments. ecoMachines turn us all into ecologists in the most operational sense of the term."[10]

10 Pasquero, C., & Poletto, M. (2008). ecoMachines for Architecture Beyond Building. Venice Architecture Biennale.

Architectural cognition

Presently, ecoLogicStudio's practice continues to question the linear association between computation and digital design, which dominates much of the contemporary architectural avant-garde, in order to propose a more evolved bio-digital paradigm where the relevance of post-natural patterns in their virtual as well as material forms acquires a cognitive potential.

The long history of computational design is dotted with those pioneers who, in different periods, have disrupted their discipline and explored different directions from the mainstream. Today, the word 'computation' needs a new definition. Its synthetic Latin origin *com-puto* (know-with) suggests alternative modes of *puto*, i.e. alternative modes of knowing. The first human machines to have been called 'computer' relied on the capability of human beings, mainly women, to compile schematic documents to be fed into the calculating machine. It was indeed a spatially distributed human–machine hybrid, a form of bio-artificial intelligence.

Meanwhile the discipline of architecture and urban design has evolved independently from such practices of bio-digital computation. So, when digitalisation of the design process occurred it finally became possible to simulate complex organisms like cities through the application of specifically customised algorithms, while the conceptual difficulty of expressing urban and architectural design problems in the form of algorithmically solvable questions is often unsurmountable. That is why ecoLogicStudio's practice has moved away from deploying design algorithms as problem-solving machines and it has evolved toward the formulation of architectural algorithms as embedded cognitive systems.

Half a century ago a group of pioneering architects mobilised computational theory and technology and succeeded not only in revising the way we draft and present architecture, but also showed architects and designers how to rethink design methods. The then newly available technologies, mainly in computational frameworks, allowed for the translation of ideological critiques and strong theoretical points into design methodologies. Computation was a means with which to prove theory. Inspired by research in cybernetics and neuroscience, architects such as Johan Fraser and scientists such as Gordon Pask became interested in the potential changes that the computer could bring to epistemological questions involving research on cognition and learning.

In this respect, the authors' research is driven by the desire to understand the process through which an organism learns, rather than uncovering the hidden organic structures of cities and picturesque villages. Computational design should evolve by establishing transversal connections across technological regimes, whereby alternative theories are developed and different ways of thinking can be constructed. Such practices will then produce novel orientations for the discipline of architecture itself. Crucially, what needs to be re-oriented through technological innovation is the thinking itself that sees technology as a means to implement norms, standards and types through the mimesis of an ideal naturalism.

We refer here to Christopher Alexander's understanding of the urban phenomenon as an actual organism[11] that is defined by means of specific part to whole relationships; this happens by means of local actions, through the naturalisation of design processes and the normativity of the model. In light of this, we should be reminded of Frédéric Migayrou's claim that:

11 Alexander, C. (1977). *A Pattern Language: Towns, Buildings, Construction*. Oxford: Oxford University Press.

"We have lost count of the number of studies that [...] endeavoured to apply the theories of complex systems and the models of biological organization to the urban arena. Behind the idea of the 'fractal city', there is often a normativeness where the models seem to stand in for the old norms of geometric space."[12]

In our epistemological quest, *Physarum polycephalum* is investigated as an example of a non-neuronal thinking entity, that has its own biases and its own workings. These functions may be no more objective or original than our own, but are certainly different. In this way, alien modes of reasoning may be introduced into design methodologies. Polycephalum does not give us what we are missing in optimality and objectivity. On the contrary, it becomes the locus for an escape towards alien forms of invention.

Audrey Dussutour of CNRS in France[13] has recently been demonstrating the cognitive capability of *Physarum polycephalum* and other microorganisms, discovering through experimental tests that cognition is not restricted to organisms holding a complex neural network or brain. Indeed, before the emergence of this recent debate on *Physarum polycephalum*'s cognitive capabilities, the slime mould had already been investigated by computer scientists for its computational skills. This type of research began in the 1980s in Japan and has been led in Europe by the work of A. Adamatzky, who has defined the computation of the *Physarum polycephalum* as 'unconventional computation'. In his *Physarum Machines: Computers from Slime Mould*, Adamatzky frames the aims of this research, arguing that:

> "unconventional computing aims to uncover novel principles of efficient information processing and computation in physical, chemical and biological systems, to develop novel non-standard algorithms and computing architectures, and also to implement conventional algorithms in non-silicon, or wet, substrates."[14]

Tero[15] and Adamtzky are perhaps the best-known researchers to have worked and produced research on the slime mould. However, Steven Shaviro has argued that their approach is all too human. He summarises it with four prevailing approaches towards the programmable aspect of slime mould:

- Observing the slime mould's behaviour to work out algorithms and code those in a computer. Isomorphism between the actual behaviour of the organism and that of the computational results is the aim.

- Treating the organism as a black box. Its phenomenal behaviour, like that of foraging, can be mobilised into solving

12 Brayer, M., & Migayrou, F. (2000). ArchiLab9 Orléans 2000. Orléans: ArchiLab.

13 Moskvitch, K. (2018, 9 July). Slime molds remember—but do they learn? *Quanta Magazine*.

14 Adamatzky, A. (2010). *Physarum Machines: Computers from Slime Mould*. London: World Scientific.

15 Tero, A., Kobayashi, R., & Nakagaki, T. (2006). *Physarum* solver: A biologically inspired method of road-network navigation. *Physica A: Statistical Mechanics and its Applications* 363, 115–119.

a series of problems. Our epistemological blind spot concerning the process that triggers the actual behaviour is not considered in the computation.

- Connecting the substrate on which the slime mould acts with electrical input and outputs. In this sense, the slime mould plays a two-fold role where it is used both as electrical conductor and as computing device.

- Treating the oscillatory behaviour of the organism as 'NOT' and 'NAND' logic gates capable of computing complex tasks.

In line with Shaviro's argument, the authors recognise the alien intelligence within the slime mould and therefore argue that the above computational approaches are all too human: human cognitive biases are imposed on the slime mould's behaviour and solutions are given within human defined problems. Instead, the value that should be recognised in the Polycephalum's behaviour is its capacity to problematise the operation of its own biases and its own capacity to discriminate its world. It is within this approach that unconventional computing with slime moulds becomes part of a truly speculative apparatus.

As exemplified by a recent project on Paris, 'GAN-Physarum', ecoLogicStudio's practice does not seek to extract efficient biological models for structuring our world. Nor does it seek to reveal any hidden structures upon which our worlds may depend. It is instead concerned with the capacity of slime mould to improvise a solution to an encounter, not in its generality, but in its singularity. This intelligence without neural networks exhibits a form of surplus value that affects and orients thinking, and that can suggest alternative modes of reasoning.

It is this surplus that affects Polycephalum's behaviour and therefore affords it the potential to escape from its stereotypical responses. Contrary to what has been argued by previous generations of computational designers, such as for example Bernard Cache,[16] we think that it is not enough to argue for non-standard modes and relations of production. We need to find in the forces of production new lines of escape. We therefore argue that unconventional computation should determine the next phase of design computation.

Steven Shaviro writes in *Discognition*:

> "Physarum thus offers a simpler instance of what, in the case of the human beings, has been called the extended mind. According to extended mind theory, cognition does not take place only in the brain, but involves the 'coupling

16 Beaucé, B., & Cache, B. (2007). *Objectile: Fast-Wood. A Brouillon Project.* Vienna: Springer.

of biological organisms and external resources.' My mental acts extend well beyond the limits of my own brain and body."[17]

If we understand slime moulds, microalgae and AI algorithms all as forms of intelligence, and we are to focus on the language that manifests itself through both material and digital patterns, then how can we, as humans and designers, enter into such a conversation? Can we remain simply as human figures in a rendered image? Or the hand that holds the pencil and the body that wears the Oculus?

What is argued here is the necessity to rethink the ways in which we describe ourselves as species and as bodies. In this context, Andy Clark, in his book *Natural-Born Cyborgs*, provides an intriguing redefinition of the self:

> "There is no self, if by self we mean some central cognitive essence that makes me who and what I am. In its place there is just the 'soft self': a rough-and-tumble, control-sharing coalition of processes—some neural, some bodily, some technological—and an ongoing drive to tell a story, to paint a picture in which 'I' am the central player. [...] The notion of a real, yet wafer-thin self is a profound mistake. It is a mistake that blinds us to our real nature and leads us to radically undervalue and misconceive the roles of context, culture, environment and technology in the constitution of individual human persons."[18]

In projects like GAN-Physarum, ecoLogicStudio celebrates how unconventional approaches to computing recognise the computational power embedded within non-human forms of intelligence that evade silicon-based microprocessors and their underlying binary logic. That is why the project constructs a new kind of drawing machine, or drawing bio-algorithm, embedding it within a territorial design apparatus capable of scanning and navigating real world territories. Originally, this apparatus was called 'Physarum Machine' and it did operate by embedding a living *Physarum polycephalum* onto a morphological substratum of a real territorial dataset that had been extracted by satellite monitoring protocols.

This apparatus is of course speculative in the sense that it embodies a future when man-made infrastructures and non-human biological systems will constitute parts of a single biotechnological whole. In this respect, it can be read as a manifesto for the scalar expansion of biotechnology to biospheric design and for the increased spatial and material articulation of global informational and energetic infrastructures as well as of the protocols of artificial intelligence managing them.

17 Shaviro, S. (2016). *Discognition*. London: Repeater Books.

18 Clark, A. (2004). *Natural-Born Cyborgs: Minds, Technologies, and the Future of Human Intelligence.* Oxford: Oxford University Press.

Polycephalum drawing

According to the Global Footprint Network in 2022 the Earth Overshoot Day landed on 28 July. In less than seven months humanity therefore exhausted Earth's budget for the whole year. For the remaining days the ecological deficit has been maintained by drawing down local resource stocks and accumulating carbon dioxide in the atmosphere; in other words, while operating in overshoot.

This mode of operation is made possible by the existence of the Urbansphere, the global apparatus of contemporary urbanity: a dense network of informational, material and energetic infrastructures that sustain our increasingly demanding metabolism while offsetting the fluctuations and deficiencies of the natural Biosphere in providing the required levels of resources in the right place at the right time. As such, living within the Urbansphere as we know it today may at times give the illusion of there being an infinite amount of available resources. The world's resources however are finite and their finitude is not normally observable directly to us as individuals—it escapes our direct perception and direct control. The paradox we face today is that while the Urbansphere has evolved into our civilisation's preferred habitat, it is also the cause of the greatest threats to our survival.

The speculative model embodied in the Physarum Machine tackles this paradox by redefining the conceptual separation of the Urbansphere from the Biosphere. The two are now so intricately connected and co-evolutionary that they cannot be separated. We therefore have proposed to define the Urbansphere as the contemporary, augmented version of the Biosphere. This observation reveals the latent potential of reframing the notion of sustainability as a design speculation, by relating the quantitative analysis regularly produced by organisations such as the Global Footprint Network to a morphological model of our urban infrastructures, all the way up in resolution to the spatial grain we inhabit and perceive in our daily experience—the resolution of architectural spaces.

The Physarum Machine embeds this relationship into a 'material algorithm', where intelligence is embodied in form and memory is perceived as a visible pattern in space.

The first instance of this experimental project was a bio-digital apparatus conceived by the authors and further developed within the Urban Morphogenesis Lab at the UCL in London. The conception of this apparatus took inspiration from the experimental method pioneered by German architect Frei Otto. As described before, Otto has studied the processes of occupation and connection of large territories by means of apparatus deploying the computational power of soap bubbles, sand, ink droplets and so on. Otto was able to compute in a completely analogical way the emergence of path systems and, in particular, to define a special category of path systems, the so-called 'minimising detour' networks. These networks are special as they produce connections between points that are optimal in terms of the energy expenditure required to connect them.[19]

Otto discovered that the patterns emerging from his experiments appeared almost everywhere in non-human as well as in human systems. Frei Otto's intuition seems more relevant today in the light of what we have described as the necessity to find models for the co-evolution of the Urbansphere within a milieu of finite resources. That is, to re-cast human infrastructures as symbiont of the Biospheric landscape. Contemporary technology offers renewed means to operate in this territory, with biotechnology allowing us not only to mimic nature but also to hack into nature's codes and augment its functionality.

Information technology in turn is enabling the emergence of global informational networks connecting all things human and non-human, living and non-living, thus implementing planetary ubiquitous computing. These influences have led to the development of the Physarum Machine, characterised by the crafted juxtaposition of a digital substratum of communication and a biological computing system.

The main driver of this purposeful hybridisation has been the attempt to explore and represent within a small-scale apparatus this notion of 'naturalness', as observed by Otto, beyond the living biological world of nature. While the Physarum Machine inherits this approach to 'natural urbanism', it also grows out of the influence of unconventional computing, a branch of computer science pioneered in the UK by Andrew Adamatzky. For Adamatzky the Physarum machine is a concept for a new kind of computing machine: "By experimentally implementing the Kolmogorov-Uspensky machine in Physarum, we prove that plasmodium of *P. polycephalum* is a general-purpose computer".[20] In this sense ecoLogicStudio's Physarum Machine research extends the paradigm of natural urbanism found in Frei Otto's work to the generative possibilities unleashed by ubiquitous computing.

The core computational system in the Physarum Machine is the *Physarum polycephalum*, or slime mould. Before going into the detailed functionalities of the apparatus it is necessary to define how the biological structure of *P. polycephalum* actually affords its computational faculties.

The slime mould is a protist, a form of single cell organism. It is however a very peculiar kind of mono-cellular creature.

19 Otto, F. (2008). *Occupying and Connecting: Thoughts on Territories and Spheres of Influence with Particular Reference to Human Settlement*. Stuttgart: Axel Menges, p.111.

20 Adamatzky, A. (2010). *Physarum Machines: Computers from Slime Mould*. London: World Scientific.

Within its extraordinary body, thousands of nuclei float in protoplasmic flows. A membrane of actine encloses the nuclei. In the plasmodium phase, when enough nutrients are available, the membrane stretches to take on virtually any morphology. The nuclei interact with each other and the environment via biochemical reactions that generate gradients of pressure which, in turn, help regulate the protoplasmic flow. The organism can sense the presence and amount of nutrients in a given place and at a given time, and react to it. This reaction is locally driven and based on multiple random interactions unaided by central planning or decision-making.

The slime mould is a special kind of biological computer that has no discrete brain but is nonetheless able to leave traces in the environment that constitute a form of spatial and distributed memory. This memory enables the slime mould to develop and optimise its behaviours. Scientists experimenting with slime mould discovered that this very simple collective organism can in fact perform extremely sophisticated tasks such as network optimisation and nutrient regulation. Under some conditions it may even anticipate future events.

So, the slime mould not only represents a simple form of bottom-up self-organisation via multiple interactions, but it is also capable of performing multiple forms of computation and optimisation without recurring to a centralised brain. This kind of optimisation is critical to its survival. Crucially for the argument made here, urban systems require similar kind of optimisation. After all, cities must adjust their networks of logistics in real time; they need to regulate the amounts of raw materials and energy they extract from each extraction/production site and are always predicting or anticipating daily, seasonal or epochal fluctuations in demand and supply.

In recent years, transport engineers and computer scientists like Tero and his colleagues in Tokyo have already recognised this opportunity to apply slime mould computation to urban engineering. They adopted slime mould as a computational model in a set of pioneering experiments developed to simulate traffic infrastructures across entire continents. In one of their best known experiments, the slime mould was capable of reproducing with surprising accuracy the network of Tokyo's Metro system.

However, the optimisation strategies through which slime mould achieved its solution differ radically from the ones adopted by the Tokyo Metro transport engineers. In fact these are unlike any planned urban systems. They are the product of emergent collective behaviour and distributed spatial memory, and in order to experiment with these strategies and develop a model for planning the Urbansphere, it is necessary to embed them directly on an augmented substratum or computational territory. This principle was tested by the authors on a new kind of bio-computational device, the Physarum Machine.

Within the Physarum Machine the slime mould grew onto a prototypical urban scenario. In the first instance, the scenario was a model of the copper mining corridor in Arizona, USA. The apparatus of the Physarum Machine materialises a hybrid, interconnected and unstable substratum for growth, which enables the slime mould to be fed with data from a territorial map and captures the behaviour of the mould as it tries to solve the feeding problem.

In this first test, three different kinds of input parameters were adopted to communicate with slime mould: substratum morphology, light fields and food source locations. Topographic and morphologic information from the real urban terrain were translated into various 3D-printed topographies, which represent morphological boundaries of matter in the real world and become the basis for the development of slime mould's spatial memory. The material therefore needs to capture the traces left by slime mould and store them in time. Initially this substratum was 3D printed in ABS plastic and coated by a thin layer of non-nutritious agar.

Gradients of light are produced by an LED matrix that the slime mould seeks to avoid as light inhibits its growth. These gradients represent areas of friction or obstacles in the real world and can change over time as a manifestation of live data streams. Finally, nutrients are inserted to sustain the slime mould's metabolism. At any given moment the mould searches for and reaches out to the nutrients, adjusts its morphology and develops its distributed intelligence while minimising the expenditure of energy to complete these tasks.

These nutrients represent sources of energy and raw material in the real world. In the Physarum Machine those are delivered by means of a customised 3D-printing machine that is capable of depositing in the exact location of study a minute quantity of nutrient. In the experiment, the slime mould is kept wet and food is dropped in the locations representing present and future mining. Such locations have been retrieved from a satellite survey where crosses mark the points of current drilling for copper as well as the distribution of testing boreholes that show sufficient amounts of copper ore to justify future drilling. The peculiar nature of the copper corridor means that copper ore is diffused in the territory at very low density. This makes open pit mining obligatory and also forces a sprawling model that leaves large areas of landscape scarred, generating a large volume of detritus that is unsuitable for use in the mining industry. It eventually becomes the sediment in large tailing ponds that further modify the original morphology of the territory. In other words, small quantities of valuable resources are scattered over a large terrain, providing the perfect testing ground for the sensitive slime mould to compute.

In the experiment the source of nutrient for the slime mould was also colour coded with natural pigments according to the specific mineral contents and status of the mining location. As the mould expands to reach out for food it forms a network and begins dissolving and distributing the nutrients along it. Since its walls are transparent, the colour in the nutrients affects the colour of the slime mould itself and allows direct observation of the spatial architecture of the slime mould's computational process. A high-resolution camera located above captures the mould's behaviour, morphology and the nuances of colour in real time.

At time zero, the slime mould is introduced in the apparatus at five specific points. From these locations it starts spreading in search of food. The slime mould's behaviour is recorded as droplets of nutrients are introduced on actual mining locations, while LED lighting patterns reproduce the main site obstacles and boundaries. Since the behaviour of the slime mould is based on an internal algorithm that optimises the expenditure of energy to reach out for nutrients, it biologically computes the relationship between the amount of nutrients available at each moment in time and their necessity to be distributed to reach all the nuclei in the cell. As the mould feeds, it visibly modifies the growth substratum producing a network of traces that improve its ability to compute. In other words, it succeeds in minimising the overall length, or volume, of its networked body to connect effectively the most profitable sources of nutrients.

However, this solution is reached only after a long searching sequence, during which time the mould tests multiple solutions or paths to connect the desired locations and evaluates their importance to the overall metabolic balance. Only after much searching and sorting the emergent process is able to select the best paths. While this may initially look like a wasteful process, it is indeed productive as it becomes clear later on in the experiment. The paths that are discarded, in fact, do remain active in the substratum as traces. These traces accumulate in time and constitute a form of distributed spatial memory. As resources are consumed the mould becomes more efficient in adjusting its morphology to rebalance its own metabolism. This adjustment is made possible by further and faster iterations of the search and sort mechanism. As the amount of traces increases these iterations become more and more effective. A moment is reached when the distributed memory allows the mould to take real-time decisions about how to optimise its morphology and rebalance resources. Slime mould at this stage of growth may even be able to anticipate the scarcity or the frequency of renewal of available resources. In other words, the harder it is to find resources the more effective slime mould becomes at finding them.

This embedded or outsourced memory transforms and incorporates the material substratum of growth into the slime mould's computational body and, as a result, its cognitive system. The outcome of this process is captured in the distribution of colours on the substratum, a true abstract bio-computational painting that encodes the morphogenesis of an emerging cognitive landscape. Transformed in time by the interaction with the living 'infrastructure' of the slime mould, this landscape depicts the encoding of a distributed memory (information) that becomes instrumental for the bottom-up bio-algorithm to optimise the delivery and exploitation of available resources (energy).

This analysis of this process has a critical significance for formulating the new vision of city that we call the Urbansphere as the Physarum Machine provides an operational model for recasting the relationship between urban infrastructure and landscape. The necessity of such reconsideration is evident in our resource-driven contemporary urbanisation and is exemplified by case studies such as that of a manufactured landscape of Arizona as well as many resource-driven cities around the globe. In that respect, the slime mould is an operational model of a self-aware infrastructural network. The computational process that delivers this awareness is internalised in the landscape that contains the resources it manages. The successful survival of the organism-computing-machine that is the slime mould body-substratum (a new unit of infrastructure-landscape) depends upon the efficiency of its computational process. In other words, this form of intelligence has evolved within a framework where the energy required to compute the problem of harvesting and exporting available resources must be less than that provided by the available resources at any given moment in time.

In subsequent experiments the embedded memory of the slime mould has been captured as drawings on various media, ranging from paper to AI datasets and swarm robots. Typically, this process involves capturing high-resolution images of the experimental terrain and developing an algorithm that recognises the morphology of the slime mould as it changes in time.

In an early set of works the virtual agents were scripted in software processing on the basis of a modified swarm intelligence model that has the ability to read and write a background map and to leave traces of movement. Those traces accumulate as the experiment proceeds and lines of movement become increasingly stable. In the emerging drawings a beautiful kind of fuzziness is produced as the edges of the drawing appear in continuous state of flux. Zooming from the territorial to architectural scale reveals more detail, but never a fixed boundary.

From a methodological point of view, this technique opens a window beyond digital and descriptive simulations. It enables a space for design where the constraints of descriptive computation are overcome and digital computation is superseded by the possibility of simulating and computing the world that surrounds us. This is possible without questioning the speculative character of analogue drawing within design explorations and without implying the superiority of generative computational models. What is described here is a bio-digital design technique that departs from any distinction between drawing and digital simulation in order to ask fundamental questions about the bases of those approaches.

Perhaps what matters the most here is to come to terms with the anthropocentrism immanent in the explorative mobilisation of drawings in architecture and urban design, and how it is limiting their operative scope.[21] This is especially true today, in a dynamic world where design problems require a broader and distributed perspective of solutions that each call for a non-anthropocentric mode of reasoning and designing. This is because, as we have seen, our impact on the Biosphere has reached every corner of planet Earth and its effects are so profound as to have exceeded our conventional understanding, so much so that we can no longer be the sole repository and arbiters of suitable (rational) solutions.

21 Pasquero, C., & Zaroukas, E. (2016). Design prototype. *AAE Conference Proceedings*. London: The Bartlett UCL, p.98.

The Physarum Machine was built and its capacities were explored precisely to offer alternative approach to the prevailing anthropic predicament in design and, more importantly, to suggest an alternative mode of operation in drawing. It explores the speculative capacities of a new type of drawing that is mobilised within the design process, and argues for the necessity to resist the substitution of living analogue models with their digital algorithmic counterparts. The *Physarum polycephalum* becomes a subject with a thousand heads, capable of dragging the paint around a canvas in its search for nutrients by navigating and abstracting a variety of landscapes in its own peculiar way. It objectifies the given substratum in various ways while also dragging the ink in unexpected directions. The drawing process and the resulting drawings become speculative objects as we begin to accept that the slime mould is a subject in its own terms.

The focus here should not be to appropriate and scale up an artefact to represent a city. The bio-painting in fact has to be drawn at the commensurate scale of the *Physarum polycephalum*. It is also not about abstracting its behaviour as a kind of minimal path algorithm, which would imply extracting a special case solution for a human-oriented problem. Rather, the main philosophical and methodological adventure here is to observe the diagrammatic capacity of the Polycephalum process of drawing.

When the problem at hand is problematised in scales and durations beyond the capacities of the human brain a new speculative horizon is constructed and the possibility for new 'What if?' scenarios is established. The capacity of the drawing to distribute existing agencies and refract new ones becomes a revisionary force for the human intellect. This refraction substitutes the all-too-human reflexive practice and re-orients human intellect and reason to scales beyond those that are readily digestible to us. In this sense, the Polycephalum drawing communicates what is otherwise impossible to communicate by human means; and the apparatus ceases to be simply a communicative and representative medium, becoming instead an object for speculation.

Thus, the proposal is not simply to connect the human eye and mind with a non-human drawing apparatus, but to radically suspend the traditional modes of drawing that in many ways are simply too human. In the Polycephalum drawing experiments, such as the more recent GAN-Physarum, the act of drawing takes place within the slime mould's capacity to drag ink and to communicate with an AI algorithm, while the connection between the human mind and the drawing apparatus is suspended. This suspension would be quickly re-appropriated by the human intellect if the living organism could be replaced by the algorithm, but computing facilitated by algorithms in digital machines cannot be equated with how 'wet' computing takes place across the slime mould's body.

To assume the opposite, i.e., to assume that one type of computation (human-algorithmic computation) is shared between different entities, is to assume that thinking is a privileged capacity exhibited by humans alone, but as we have seen, thinking also takes place in slime mould through a peculiar perspective that enables it to construct its own kind of 'computing algorithm'. After all, the Latin origin of the word compute is in *com* (together) and *puto* (thinking). *Physarum polycephalum* abstracts and constructs a very different computational world from that which we normally associate with 'computing', but still 'computes' nonetheless.

To conclude, if drawings are to have any future in architecture, a practice that is becoming increasingly dominated by protocols of communication between machines and forms of artificial intelligence that make any human readable notational system obsolete, it must convey to our world traces of an alien view, to inform, revise, update and reorientate our human intellect.

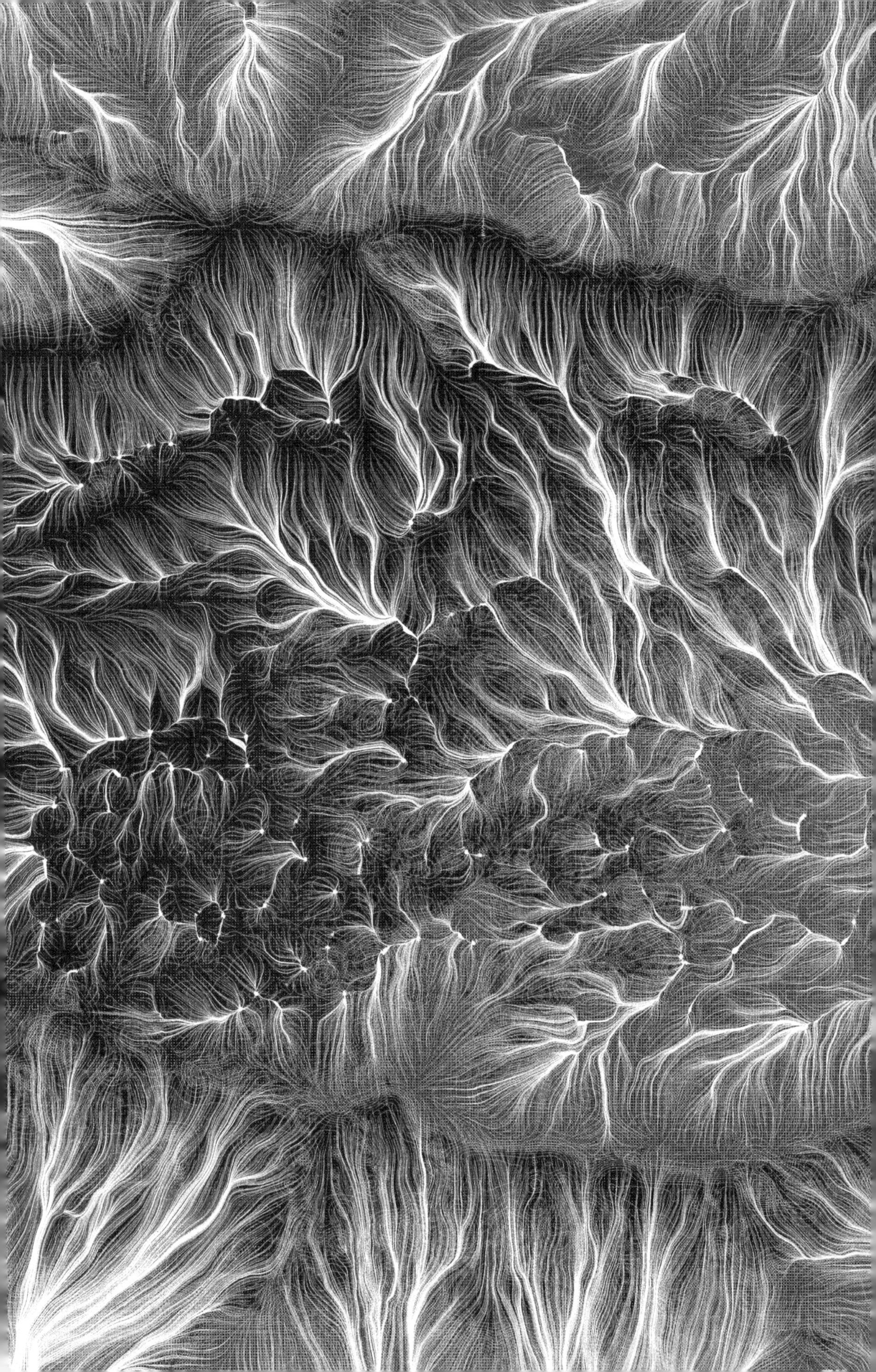

Projects 2.01

SYNTHETIC LANDSCAPES

2.01.01 Les Jardins Fluviaux de la Loire 254

2.01.02 Solana Open Aviary 266

2.01.03 The Anthropocene Island project 282

2.01.04 Biocoenosis Nest: A city woven by collective intelligence 294

2.01.05 GAN.OS 101 298

♪♪ UNESCO recommends preserving the Loire as a 'wild river'. We propose to turn such turbulent wilderness into a new model of urbanity for the city of Orléans, a new public fluvial garden located in the Loire's riverbed.

2.01.01

Les Jardins Fluviaux de la Loire

This project is a landscape and urban project commissioned for the Orléans Biennale 2017, entitled 'Les Jardins Fluviaux de la Loire – Turbulent Urbanity'. It investigates the possible future of the Loire as a wild river and its controversial relationship with the city of Orléans through a multi-scalar approach spanning from satellite data to microscopic observations.

Methodologically this project expands on the exploration of bio-computational interfaces applied to the landscape as well as at the territorial scale. It follows on a line of enquiry previously investigated in the 'Synthetic Crystallisation' project and which will be expanded here in terms of its strategic and design outputs.

The Loire valley is a 'cultural landscape' synthesising knowledge belonging to different disciplinary domains: architecture, landscape design, agriculture, hydrology, geology, social sciences, economics and so on. The importance of that definition is reinforced by the recognition it has received from UNESCO, which underlines the branding and economic value that the Loire topographic region embodies. While France, as a nation state, is losing its importance in the contemporary global order and Orléans, a small city, does not appear on the global map of emerging metropolises, the Loire Valley possesses sufficient character and diversity to be a recognisable attractor. It is therefore sensible for the city of Orléans to frame its new metropolitan status in relationship to the Loire as a topographic region.

However, a key factor in the current definition of the Loire valley as a cultural landscape is the preservation of the Loire as a 'wild river', which implies avoiding urban infrastructures that may fundamentally alter its course or deplete its biodiversity. Therefore it may seem contradictory to seek to establish a stronger urban dimension for a domain whose value is measured in terms of its wildness. We would argue however that engaging such a contradiction may offer an opportunity for both the Loire valley as a whole and the city of Orléans to propose a new model of urbanity, one that exists outside the ideological struggle of natural conservation versus urban development.

One of the most problematic aspects in the relationship between a 'wild river' and a substantial urban development today is the aspect of risk management. Wild rivers, and the Loire is no exception, have a non-linear behaviour that can become highly unpredictable and fluctuates dramatically from season to season and from year to year. Meanwhile, cities like Orléans demand high levels of control over unpredictable and potentially destructive natural phenomena such as flooding.

Friction—city vs. river

In the past, the management of this elastic relationship has implied not only the building of defensive structures, but, especially the practice of constant monitoring and intervention. Such monitoring was a direct function of multiple activities that were taking place in and on the river, such as commercial navigation, and that did involve a large section of the population and local institutions. Today the river has become more of a backdrop to urban activity than a medium to facilitate its development. Urban life is only indirectly affected by the river and its course, and as such both local institutions and the community at large have happily delegated its management to specific organisms.

These management protocols have therefore become obscure to the point that very few individuals have any knowledge of them. As such their general understanding has become ideological: everyone expects risk managers to simply minimise or erase risk, but the 'How to?' is little known.

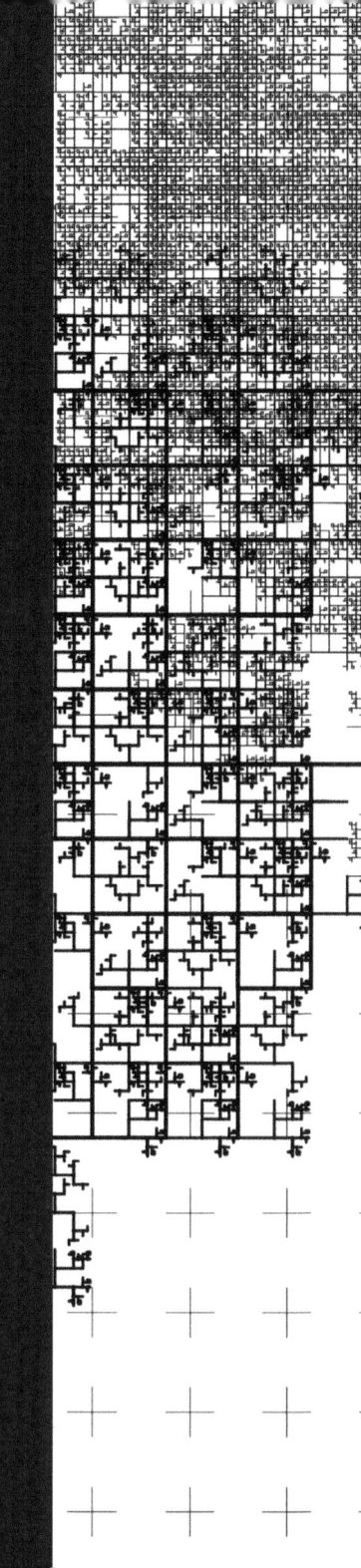

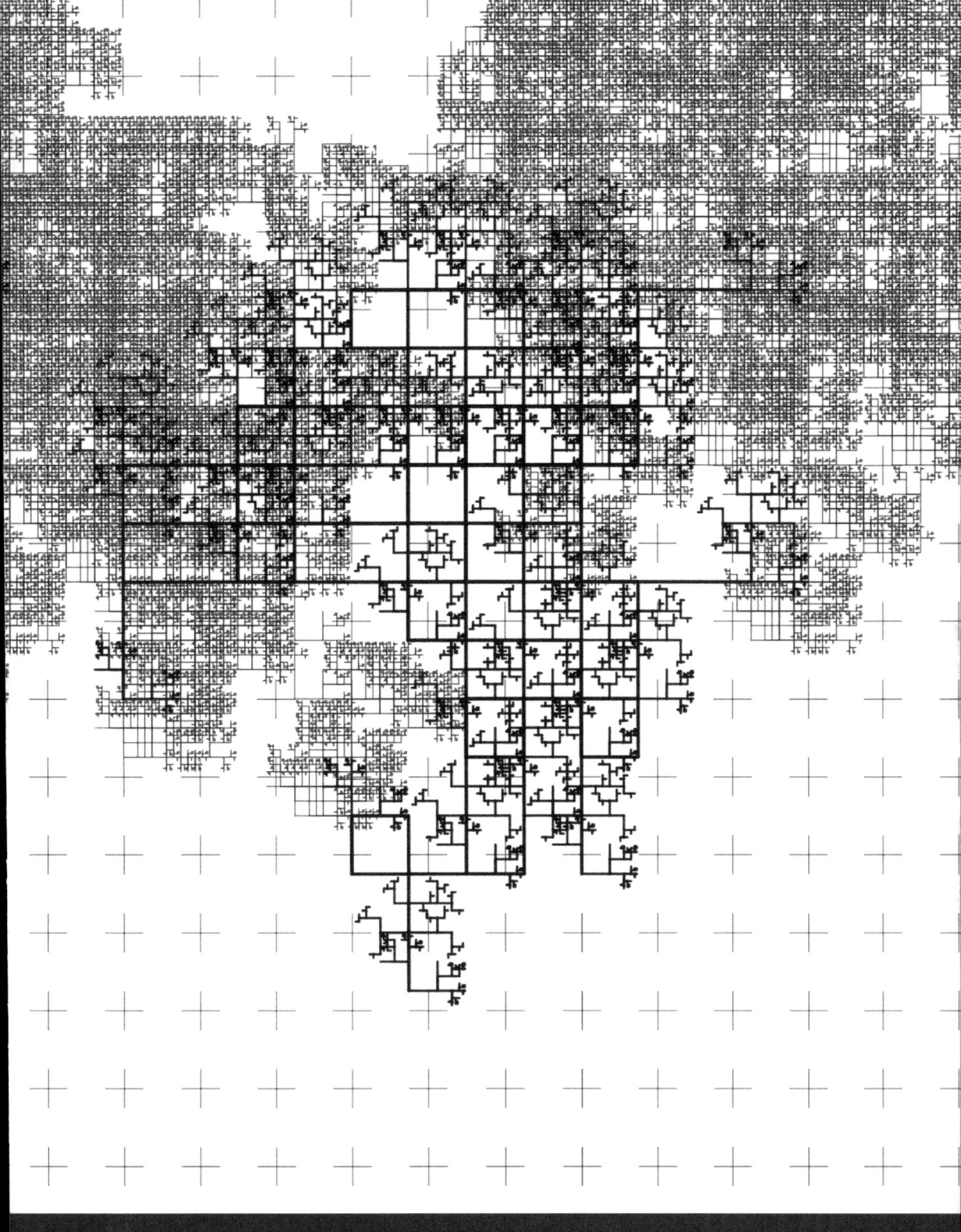

Cover page: 3D-printed model of the Garden of Accretion at Pont Georges V in Orléans.

This page: Algorithmic simulation highlighting the systemic nature of the proposed Garden of Sedimentation. The plan visualises the crystallisation patterns forming within the Loire riverbed in Orléans.

View of the project site highlighting the turbulent flow of the Loire near Pont Georges V in Orléans.

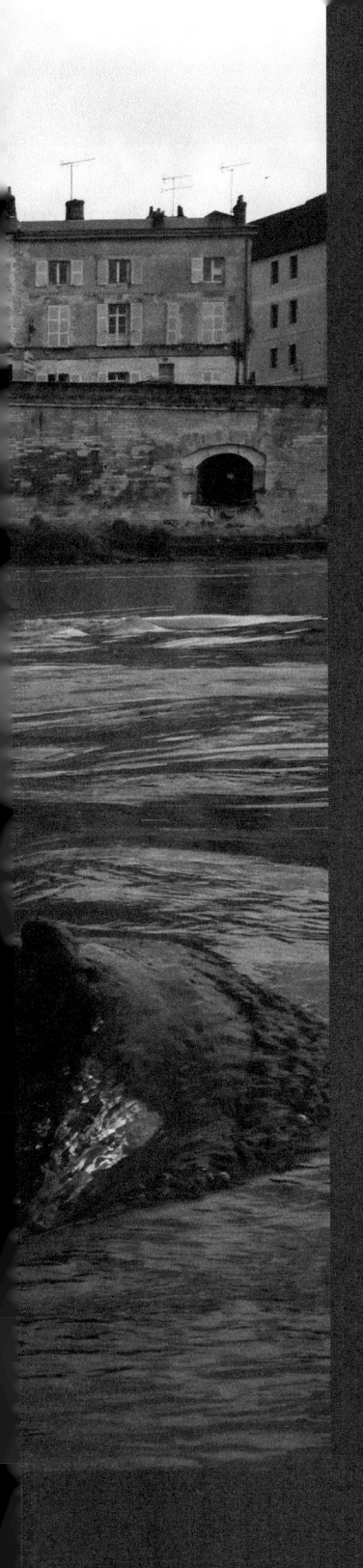

Sedimentation

One of the key potentials offered by the status of the Loire as a wild river is that no permanent canalisation or dam has been built along its lower course. In the city of Orléans in particular, the river is bound by a soft flood barrier that still leaves plenty of room for the river to expand and contract. Only a longitudinal dam has been built within the riverbed with the function of clearing a navigable edge towards the old city. As such the river still manifests its non-linearity which can be observed in the turbulent patterns of flow as well as in the sedimentation of islands of various forms and dimensions; such geomorphodynamic formations are rather accelerated. An early satellite analysis shows the movement is seasonal and the riverbed morphology is being redefined over a matter of months.

This positive feedback is contrasted and regulated by a series of other emergent biotic and a-biotic processes, some also highly influenced by human action, albeit involuntarily. A critical one is the spontaneous growth of vegetation on river islands. As rural areas proliferate along the valley with cultures such as grape becoming a prominent symbol of the cultural landscape of the Loire, the soil of the riverbed has become saturated with fertilising substances that rain washes into the river itself. Consequently, the Loire is today rich in phosphates and nitrates that contribute to the rapid growth of small and large vegetation in the riverbed and on the sediment islands. This process has the effect of consolidating the riverbed and adding biomass. The balance between erosion and sedimentation is highly unstable and can rapidly escalate.

Interestingly, and somewhat paradoxically, this emergent anthropic landscape of sedimentary gardens is collectively conceptualised as a testament of wilderness, triggering conservationist instincts which have the effect of tipping the balance of forces towards a greener riverbed. However, it is important to notice that the notion of self-regulation in dynamical systems like wild rivers does not imply an ideal and balanced state of greening. On the contrary, episodes of wild reorganisation and even destruction of sedimentary landscapes are critical to the ability of the river to self-regulate. So how can we reconcile this need for movement and morphological self-organisation with the imperative of a modern city built on the ideology of balance and stability?

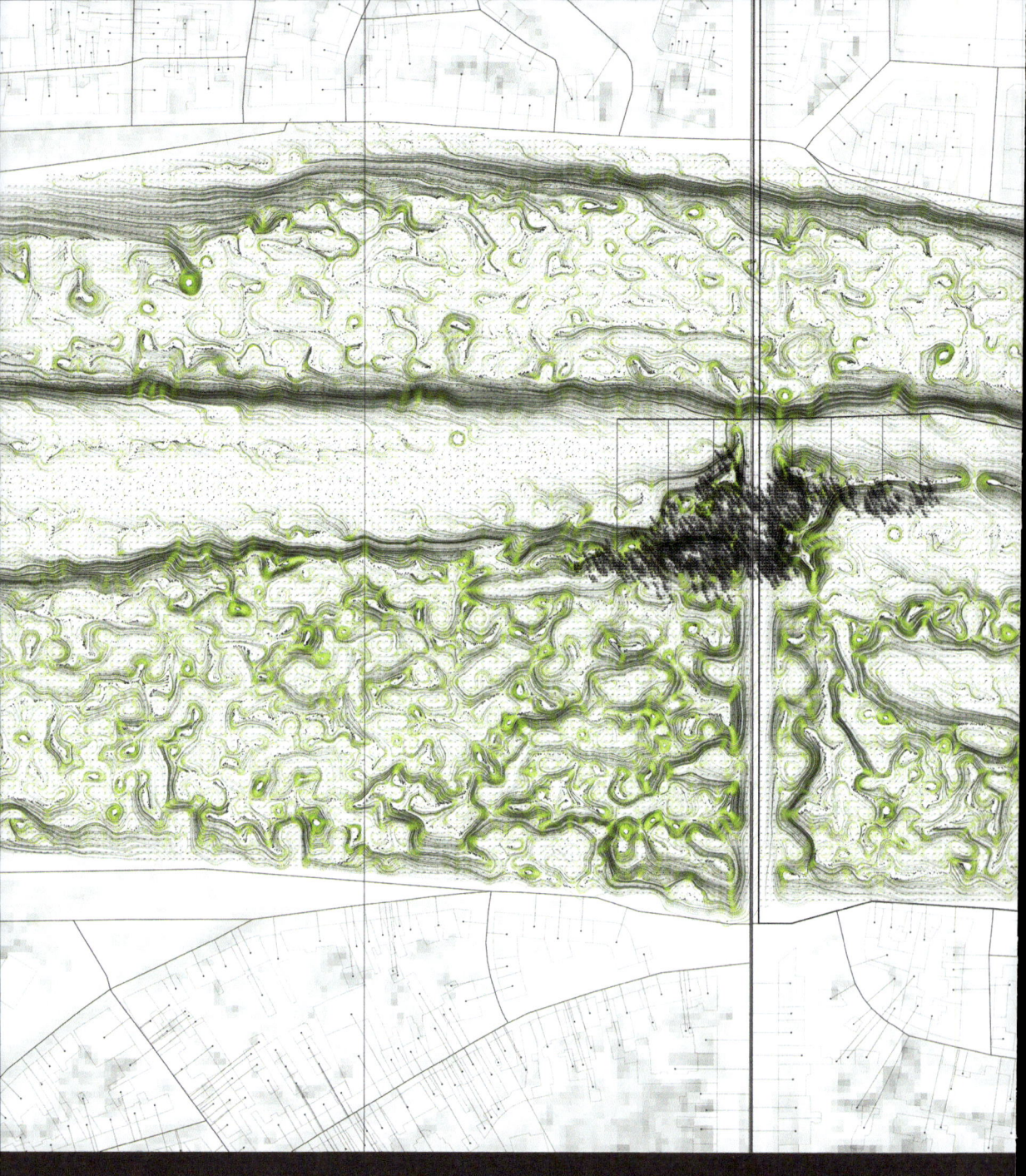

Territorial masterplan of the Jardins Fluviaux de la Loire, in the centre of Orléans. The plan shows the simulated location of the three proposed gardens, Garden of Accretions, Garden of Sedimentation and Garden of Entanglement.

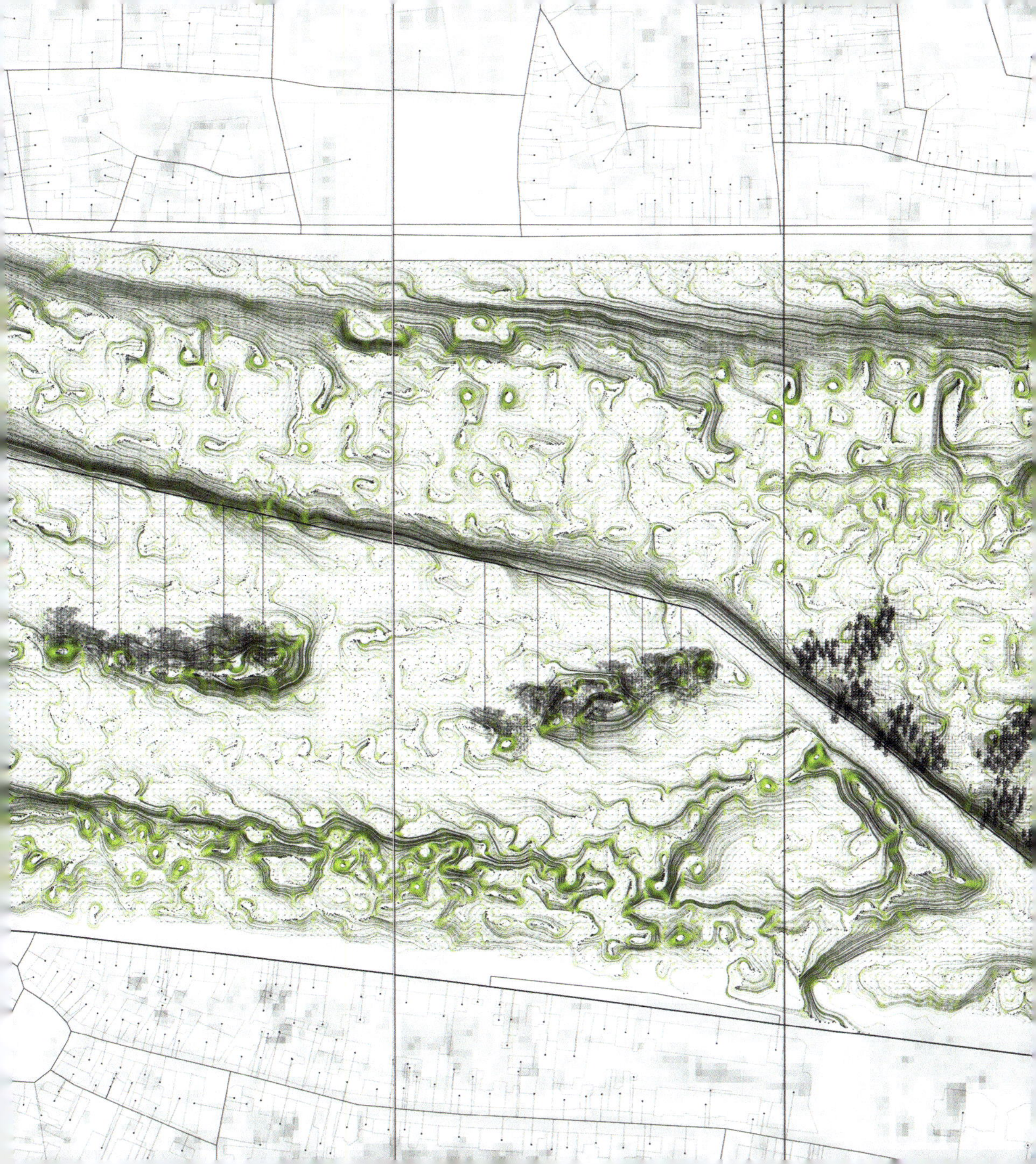

Natural crystallisations in the Loire's riverbed in winter 2017, Orléans.

Management protocols

It is therefore left to risk managers to decide when and how to manipulate some of those formations in order to minimise the risk of catastrophic events. The rationale behind such design processes, which result in the remodelling of sections of the riverbed, is obscure in the sense that it is not shared. As such it cannot be urban or evolve a new urbanity in co-evolution with the river Loire.

In reality a very large number of processes contribute to the complex dynamics of the Loire and its effect on the city. For instance, its microclimatic effect is evident. As water evaporates, especially next to more turbulent areas, haze, fog, dew and ice form. Such formations vary greatly and in winter for instance we observed the formation of large ice crystals. Ice and crystals affect the sedimentation patterns of the river, but also directly index its substratum, revealing the complexity and differentiation of the riverbed itself beyond the simple presence of vegetation. Microclimatic effects are ephemeral, but their observation and affects are critical to the dynamic of the river. With the aid of satellite and other digital monitoring techniques it is today possible to register and index these emerging patterns, visible manifestations of the fundamental geomorphodynamic processes underpinning the life and behaviour of the river Loire as it flows within the urban terrain of Orléans.

The Gardens

With the aid of satellite monitoring techniques this project registers and indexes emerging fluvial patterns, visible manifestations of the fundamental geomorphodynamic processes underpinning the life and behaviour of the river Loire as it flows within the urban terrain of Orléans.

We propose to make use of such techniques and turn them into urban design tools. To this purpose we have laid a computational grid over the river. The grid operates as a sieve, defining the resolution of information to be extrapolated from the monitoring of the river's behaviour. The indexical grid is computational but the intelligence it mobilises is material and can be deposited as a set of physical gardens, 'Les Jardins Fluviaux de la Loire'.

At the core of this project is a proposal for three urban gardens, grown within the bed of the river Loire through the introduction of a substratum in the riverbed itself, physically connected to Pont Georges V and Le Duit Saint Charles. The substratum is the result of a set of digital simulations, deploying L-system-inspired growth models.

The three gardens differ in degree as they engage the river at different resolutions. The 'Garden of Entanglement', located upstream of Le Duit, is designed to trap macroscopic material, like floating branches and plastic waste, cleaning the river and preventing its obstruction during rain storms. The 'Garden of Sedimentation', located just downstream of Le Duit, traps transported soil and rock particles, managing the formation of islands and beaches. The 'Garden of Accretion', located at the intersection of Le Duit and Pont Georges V, enables the accretion of mineral and biotic molecules over its substratum. It has water purification potential and embodies a first new link between urban infrastructure and the river.

By providing direct and safe access to the river, the Garden of Accretion promotes the emergence of a new urban practice, fluvial gardening, the collective care and management of Orléans's most precious urban landscape. It is a projective mechanism of future city making.

Indexical protocols

Our project argues for the deployment of a computational grid over a specific portion of the river Loire facing the historical centre of Orléans. The grid will operate as a sieve, defining the resolution of the grain of information we extrapolate from the monitoring of that portion of river. We will in other words index the river in space and time and at a certain resolution, and render such information visible in a set of 'operational field' drawings. As these drawings enter the public domain they will become the basis for the invention of a new collective protocol of management of the river Loire.

The operational fields themselves are purely computational entities, but they can be materialised through the application of specific media. In our case we will deploy analogue media that have a direct material effect on the sedimentation patterns of the river as they directly affect the aggregation of transported rock and soil particles . As such their deployment will not only serve as a materialisation tool but at the same time it will provide a means of direct intervention into the riverbed. Indexing will become an act of design. In this case, the informational files will also inform local techniques of mineral deposition and crystallisation, microbial aggregation and bacterial weaving.

Artificial crystallisations on a digitally fabricated substratum.
The model simulates the physical and structural evolution of the
proposed Garden of Accretions.

An augmented ornithological park where people can study, experience and affect the co-evolution of humans and birds. Or, literally, an aviary without a net.

2.01.02

Solana Open Aviary

In the now established tradition of practice-driven research, ecoLogicStudio's experimental design activity has progressed under the impulse of direct engagement with real contexts and design briefs as well as through collaboration with academic and design institutions. One extraordinary project that the firm became involved in, in 2016, was the transformation of the Solana Ulcinj in Montenegro. The project is the brainchild of architect Diana Vucinic of the Montenegro Ministry of Tourism and Sustainable Development, with Professor Bart Lootsma (University of Innsbruck) and Katharina Weinberger (architecturaltheory.eu) who were curators of the Montenegro Pavilion at the Venice Architecture Biennale 2016 where it was first presented. The Pavilion featured four project proposals, one of which—the Solana Open Aviary—was realised by ecoLogicStudio.

With a surface area of 14.9 square kilometres (5.8 square miles), Solana is one of the largest salt marshes in the Mediterranean region, a man-made landscape created in the late 1920s which has come to be regarded as an exceptional biotope of local, national and global importance. Since the decline of salt production as a viable activity in the Mediterranean, it has become critical to find new models to reprogramme coastal artificial landscapes such as this, by enhancing their touristic and cultural value as ornithological parks and articulating their new-found interest as marine resorts. The Solana Ulcinj project was conceived as a challenge to "come up with new proposals for saving the important ecological condition of the Solana Ulcinj and the unique cultural qualities of its landscape, while at the same time enabling and regulating economic interests in the area", as Professors Lootsma and Weinberger explained in their brief.

Given the complexity of re-programming of such a large territory and under the premises of practice-based research, ecoLogicStudio (Marco Poletto, Claudia Pasquero with the assistance of Terezia Greskova and Vlad Daraban) teamed up with three research partners: the Urban Morphogenesis Lab (research team: Claudia Pasquero, Maj Plemenitas and Stuart Maggs) at the Bartlett School of Architecture, University College London (UCL); the European Space Agency (ESA) research group based in Rome (led by director of research, Pier Giorgio Marchetti); and Aarhus School of Architecture (lead researcher: Marco Poletto).

Cover page: Aerial view of the proposed artificial landscape for Solana Open Aviary presented at the Montenegro Pavilion at the Venice Biennale 2016.

This page: Normalised Difference Water Index Operational Plan covering an area of 25x25km at a resolution of 10m. The plan is based on Level 1 data from the ESA Sentinel-2 satellite and it describes the territory in terms of degrees of wetness. The wet zones are rendered in purple while the dry zones are rendered in dark green.

Habitat Gradients Operational Plan covering over an area of 25x25km at a resolution of 10m. The plan describes the territory in terms of degrees of vegetation density and chlorophyll content. This is made possible by the multiple sensors available on the Sentinel-2 in the near infrared spectrum. The densest zones are rendered in purple while the sparsest zones are rendered in green. Birds' habitats are redescribed as finely articulated gradients of transition and difference.

Responding to the specificities of a brief so ingrained with notions of artificial ecology and infrastructural landscape, the work of ecoLogicStudio began by uncovering, measuring and evaluating the latent potentials in the Bojana-Buna Delta region, which includes the area occupied by Solana. The ambition was to propose a method and related workflow to productively develop a series of pilot projects capable of catalytic action at multiple scales. ecoLogicStudio's intervention focused on the generation of an urban co-action plan, with a logic of co-evolution of natural and artificial systems.

The plan frames the site through managerial zones at different resolutions: molecular, focusing on photosynthetic bacterial ecologies in the Solana's salt crust and brine; architectural, looking at the Solana's local ecology in relationship with its machines and water flow-regulating devices; and global, concerning networks and behaviours of migrating birds. It became clear from this analysis, furthered through site visits in spring 2016, that a social disconnection between urban development and the understanding of the local landscape was taking place, with evident negative repercussions for both realms. As a consequence, ecoLogicStudio's proposal developed into the notion of 'Open Aviary', as a means to stimulate reconnection and intensify co-action between socioeconomic groups and their immediate surrounding landscape.

Solana Open Aviary masterplan Covering an area of 10 x 5km. The masterplan is articulated in four operational layers each amplifying the Level 1 information provided in near real time by the ESA satellite Sentinel-2. The four layers are:

1. A network of transportation infrastructures (water, energy, vehicular and pedestrian traffic) connecting Solana with the surrounding urban systems and seawater management infrastructures.

2. The field of airborne birds' trajectories connecting local feeding, nesting and resting habitats. This operates in response to the infrared scanning of the territory revealing vegetation density and related chlorophyll concentration on the ground.

3. A layered substratum of mineral accretions, resulting from the intensification of the turbulent flows of brackish waters and the seasonal cycles of evaporation and rain. This includes a zone of robotic activity.

4. Clusters of architectural prototypes, the Open Aviaries. Each prototype hybridises multiple architectural programs including research, education, sport-leisure and healthcare. The five main prototypes are: Bird Therapy, Ornithological Campus, Micro-biological Spa, Biocatalyst power station, Bird-watching Network.

The key design concept of the Open Aviary is inspired by the possibility of digitally tracking and simulating global migratory fluxes of birds and promoting the emergence of a new concept of natural reserve as a boundless, open and networked man-made ecosystem. The conditions of the artificial territory formerly occupied by the salt production plant make the context ideal for the actualisation of this concept, into what would be the world's first open aviary, an augmented ornithological park where people can study, experience and affect the co-evolution of humans and birds.

On site the proposed Open Aviary will be literally an aviary without a net, where birds and humans explore close interaction without being forcefully enclosed in a confined envelope. This is made possible by a carefully designed workflow managing the flows of information, matter and energy provided by the deployment of digital bird-tracking technologies, high-resolution satellite earth monitoring and robotic land 3D scanning/sculpting.

ecoLogicStudio's integrated design of digital technologies and landscape ecology confers multiple meanings on the word 'open', both in the physical and virtual realms: as an open-source ornithological database, as open networks of migratory sites, and as an open-systems habitat.

View of the Solana Ulcinj in winter 2015. Detail of a disused evaporation pool with salt crystallisations and wild plant ecologies.

The 3D-printed model of the proposed artificial landscape of the Solana Open Aviary. The morphology is simulated with an agent-based edge-tracing algorithm which reads ESA infrared maps as an input and tests emerging landscape tectonics produced in time by natural mineral accretion modulated by on-site robotic extrusion.

Through trans-disciplinary design the project embraces the implications of its concept at all scales, from the intercontinental to the molecular. Firstly, the biopolitics of the Open Aviary is a biopolitical simulation at the intercontinental scale describing how habitats across different countries are in fact part of a single project when it comes to preserving the global bird population and their complex migrating behaviours.

Secondly, the operational field of the Open Aviary is facilitated by a satellite-enabled survey at the regional scale, carried out in partnership with ESA. Through the eye of Sentinel-2, ESA's new high-resolution earth-monitoring satellite, the Open Aviary embodies never-before-seen detailed scanning of biochemical processes on the ground and water, revealing a landscape that is inextricably the product of the combined action of human agency and technology with local biological life.

The third level of the project's design is the tectonic of the Open Aviary, a robotically fabricated artificial landscape that enables a new life for the salt marsh and its infrastructure in the bio-digital age. Using local material, such as the unique black clay mud and the salt crystals, natural mineral accretion is iteratively 3D scanned, accelerated and articulated to evolve a highly differentiated landscape. Its articulation makes it capable of attracting a wider variety of bird species present in the region and accommodate multiple architectural programmes, from research to sport, leisure and healthcare.

To articulate these three levels of the project, ecoLogicStudio collaborated very closely with its main research partners. Combining forces with ESA in Rome we defined a workflow enabling the team to design with near real-time data from their new Sentinel-2 satellite. ESA provided the lab with a set of virtual machines from which Level 1 data were downloaded with a resolution of 10x10 metres; a dedicated software installed on the virtual machines enabled the processing of those datasets with various algorithms into a set of false colour images, called 'infrared', 'normalised water' and 'vegetation' indexes.

Such images and their datasets are input into ecoLogicStudio's design simulation workflow, mainly developed within the Grasshopper platform from McNeel. Such workflow enables also direct communication with machines for both 3D printing and for on-site robotic actuation.

The sensors on the new Sentinel-2 satellite are capable of detecting biochemical processes in the project site as well as pollutants, and of distinguishing between multiple species of plants and soils. Such unprecedented level of detail has enabled a redefinition of the notion of habitat, as depicted by the operational field maps: an operational concept of habitat is essential to the idea of the Open Aviary, where the terrain becomes co-evolutionary with its inhabiting forces, be they human or non-human. The result is what ecoLogicStudio defines as in-human landscape or architecture, co-produced by human and non-human forces and therefore emerging as a new ontological object.

This is a shift in perspective enabled by the specific research strategy that ecoLogicStudio has adopted in the conception of robotically enabled landscapes; natural processes such as mineral accretion, typical of wetland, salt lakes and salt marshes, are accelerated and digitally manipulated to enable the emergence of novel morphologies. These bio-digital material systems conjure scenarios for the manufacture of an adaptive substratum for the Open Aviary to host both human and aviary programmes in the near future. Ultimately a real-time feedback between remote sensing and on-site intervention in the landscape determines the morphogenesis of the Solana Open Aviary. It is shaped by the dialogue between natural and artificial, local and global ecologies.

3D-printed nylon model of an Open Aviary. This prototype was laser sintered in layers of 10 microns from nylon powder giving incredible detail and flexibility to the fibres. The morphology is the result of swarm behavior simulations inspired by the flocking of birds over the Salina.

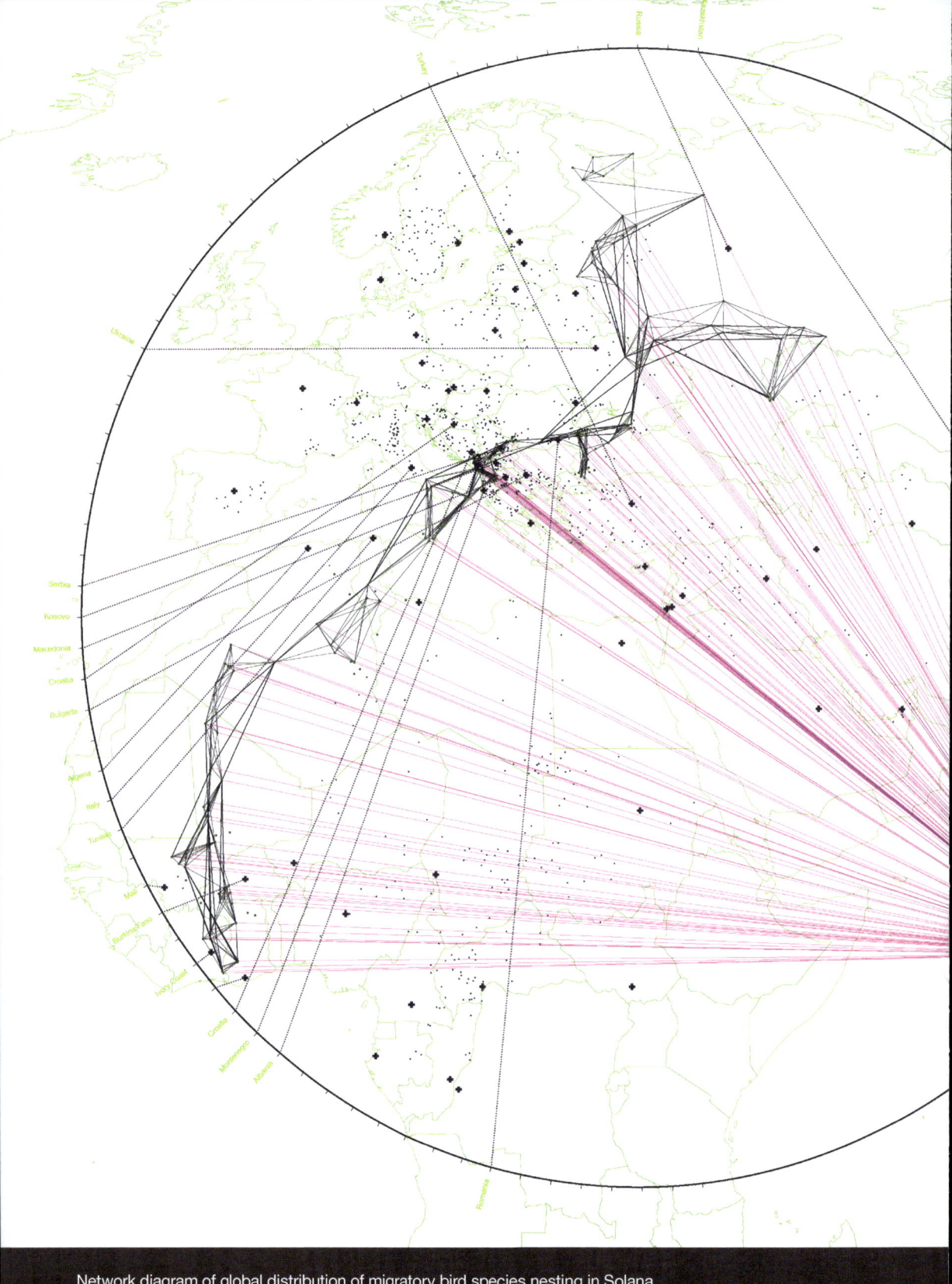

Network diagram of global distribution of migratory bird species nesting in Solana Open Aviary. The diagram highlights the geopolitical dimension of the project.

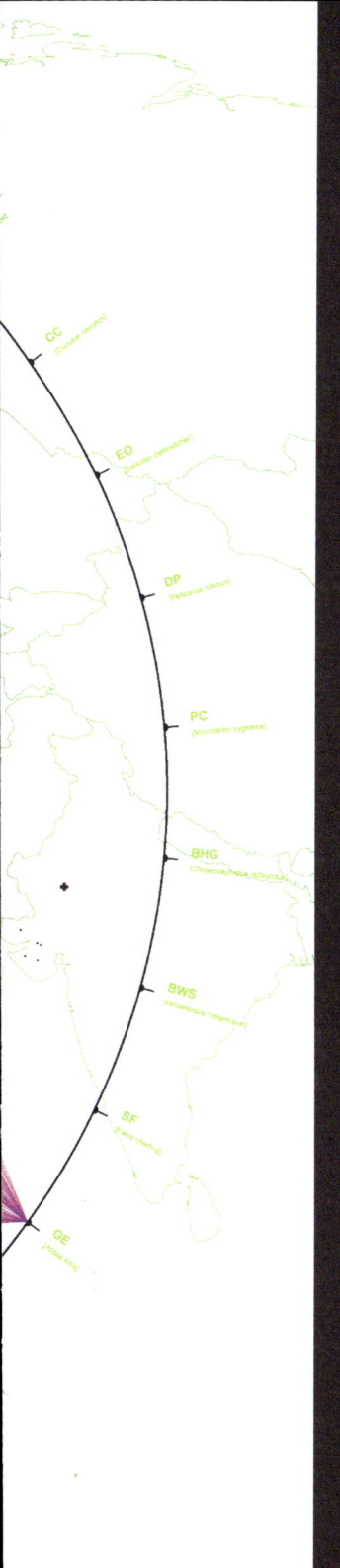

This strategy actualises the interfacing of multiple surveillance mechanisms and proposes their fruition as open-source databases of knowledge. Satellite monitoring, drone 3D scanning and ornithologists' fieldwork are interfaced and inform each other, promoting a new form of citizen science whereby both locals and tourists become actively engaged in the process of understanding as well as transforming the Solana Open Aviary.

On reflection on this case study we have come to believe that such an approach could constitute a powerful response to some of the most urgent problems affecting the site, such as poaching, which is currently driving down the number of protected birds in the area, and uncontrolled development occurring primarily on the strip of land dividing the Solana from the Mediterranean Sea. But of course these are issues affecting the whole Urbansphere, therefore the strategies adopted here have broad applicability.

The protection and conservation of the Solana Ulcinj's cultural identity as an infrastructural landscape and its ecosystemic value as an ornithological park are achieved through the hyper-articulation of its boundaries. This is illustrated architecturally by the 3D-printed study models exhibited at the Montenegro Pavilion during the Venice Biennale of Architecture 2016. Increased habitat articulation, informational exchange and network connectivity turn Open Aviary into a more resilient system, one that is inherently adaptive and receptive to future evolution. From this point of view Solana Open Aviary is a prototypical case study of a strategy for the spatial articulation of the Urbansphere. This realisation is informing our current project in Tallinn and has influenced much of our thinking.

Tallinn Wet City is a symbiotic anti-city. It co-evolves present day Tallinn to redefine its entire urban metabolism.

2.01.03

The Anthropocene Island project

In 2017 ecoLogicStudio curated the Estonian Biennale of Architecture entitled 'BioTallinn'. As part of the main curated exhibition we developed an urban proposal for the future of Tallinn, with a specific focus on the Paljassaare Peninsula. This envisioned a new urban centre emerging from the processing of the city's waste, a symbiotic anti-city that would co-evolve present day Tallinn and redefine its urban metabolism. This proposal is now the subject of further development, as ecoLogicStudio was invited to be part of a research team that is currently working on the blue-green masterplan for Tallinn. The current project is titled the 'City Unfinished' and involves the city council, a developer and researchers from the Estonian Academy of Arts.

When first approaching the project site we realised that future visions for Paljassaare Peninsula are shaped by two forms of conflicting ideology: environmentalism (the Peninsula is included in the Natura2000 network of protected areas) that strives to maintain the site in a state of illusionary wilderness; and commercial development that envisions its urbanisation into an idealised green city. The two narratives, while apparently opposite in their intents, are both ideologically conservative in their reading and understanding of the true nature of the site. ecoLogicStudio's proposal challenges such conservative sentiments with a masterplan intended to promote a new urban morphogenesis, whereby Tallinn's actual urban wastewater infrastructure is made to affect the biotic substratum of the peninsula.

The resulting 'contamination' becomes a morphogenetic force, inducing an artificial hyper-articulation of the landscape and its living systems which evolve into an urban digestive apparatus. Pathogens are re-metabolised, diluted or captured by augmented ecosystems; and infrastructural networks thicken into filtering surfaces, which in turn fold into convoluted epidermises populated by a large amount of inhabitable bio-reactor cells—the 'Anthropocene Islands' of Paljassaare.

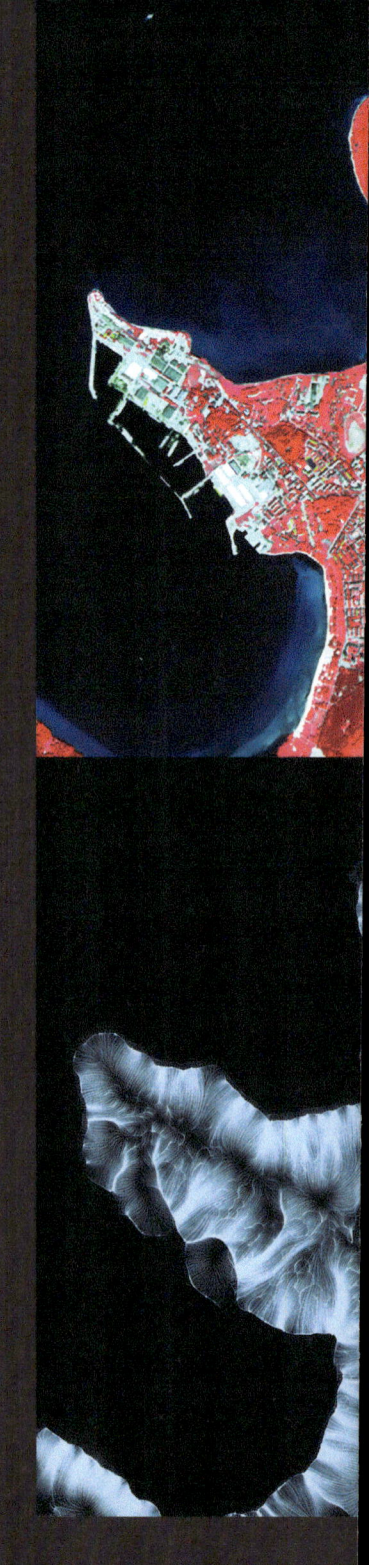

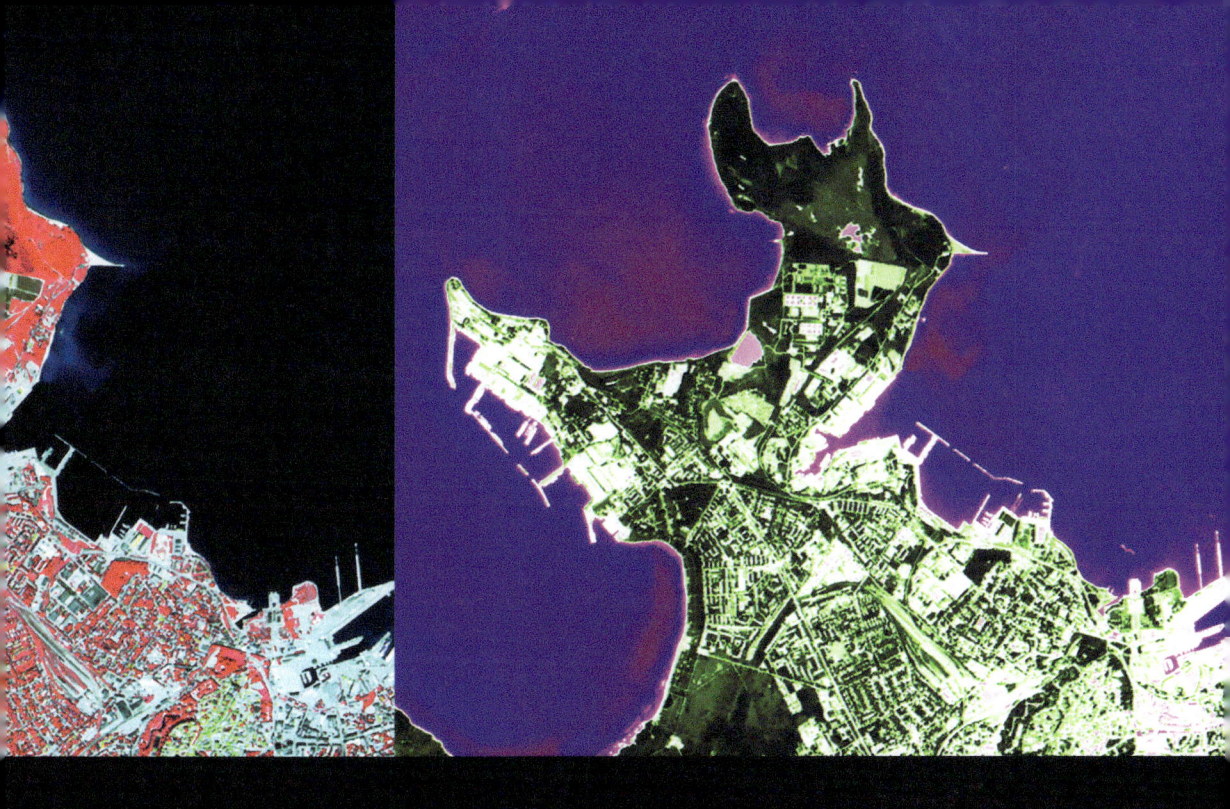

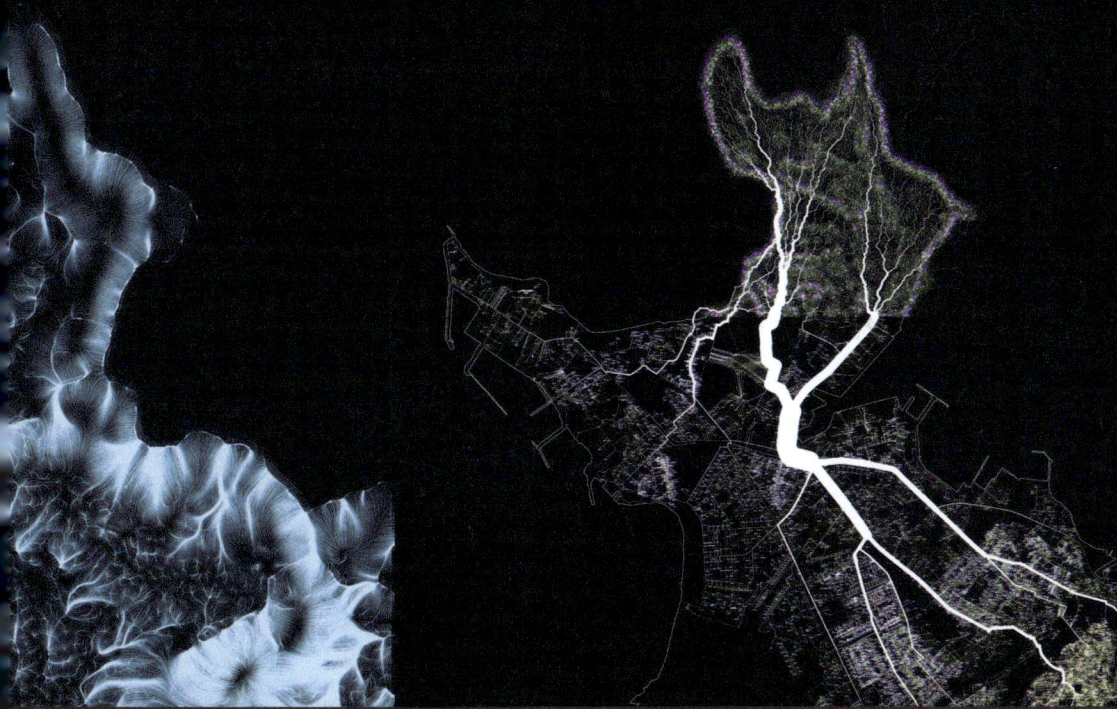

Cover page: Aerial view of the Anthropocene Island project, on the peninsula Paljassaare in Tallinn.

This page: Tallinn Wet City. Four-layered algorithmic satellite analysis of Tallinn. The four layers are (from top left to bottom right):
1. Infrared analysis of vegetated systems
2. Normalised difference water index
3. Rain and water flow patterns of the aquatic layer
4. An urban wastewater network

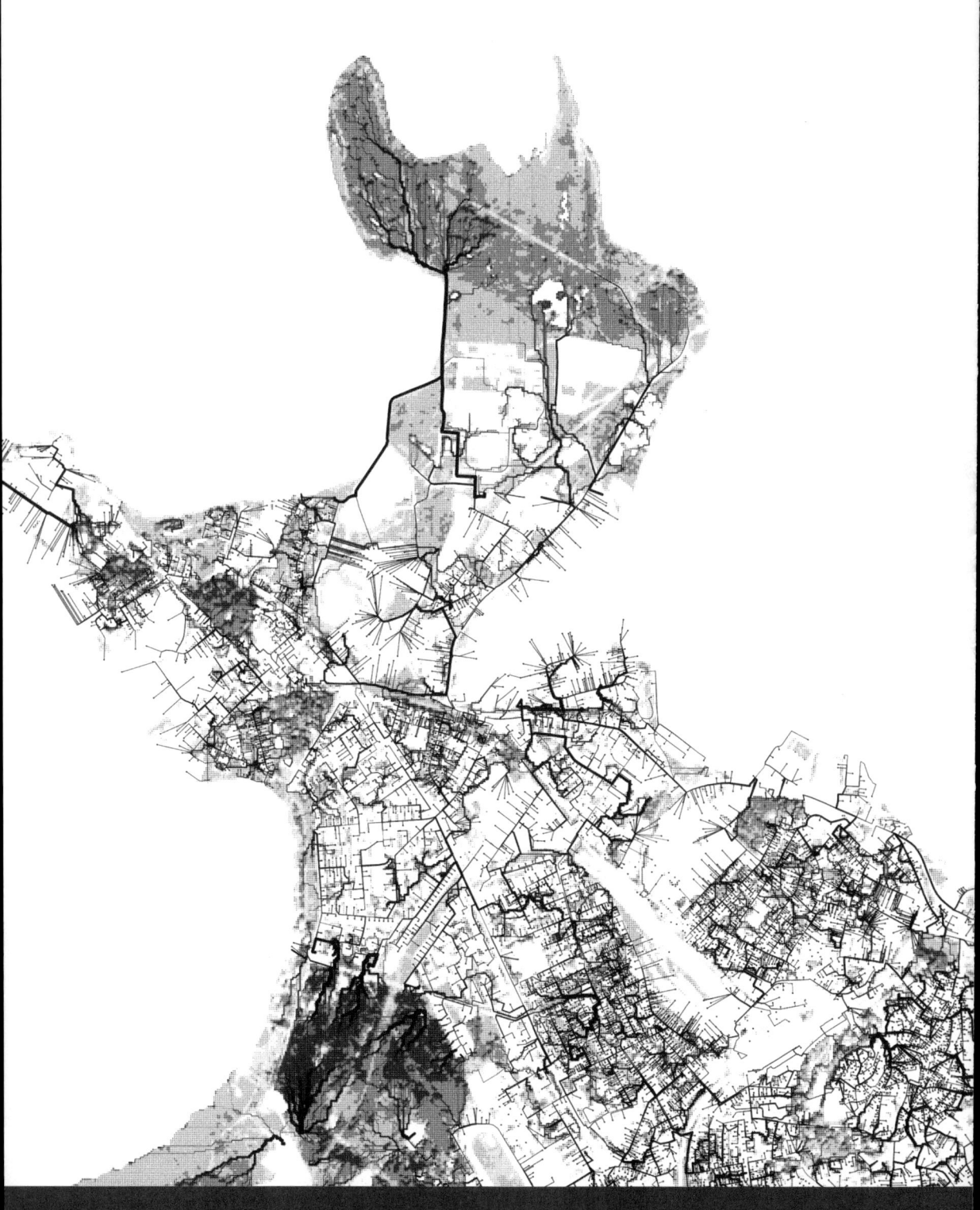

Proposed blue-green masterplan for the Talllin Wet City project. Green and built areas are treated as a co-evolving network. Links with higher biological traffic are thickened in the drawing and become prime sites for bio-architectural interventions.

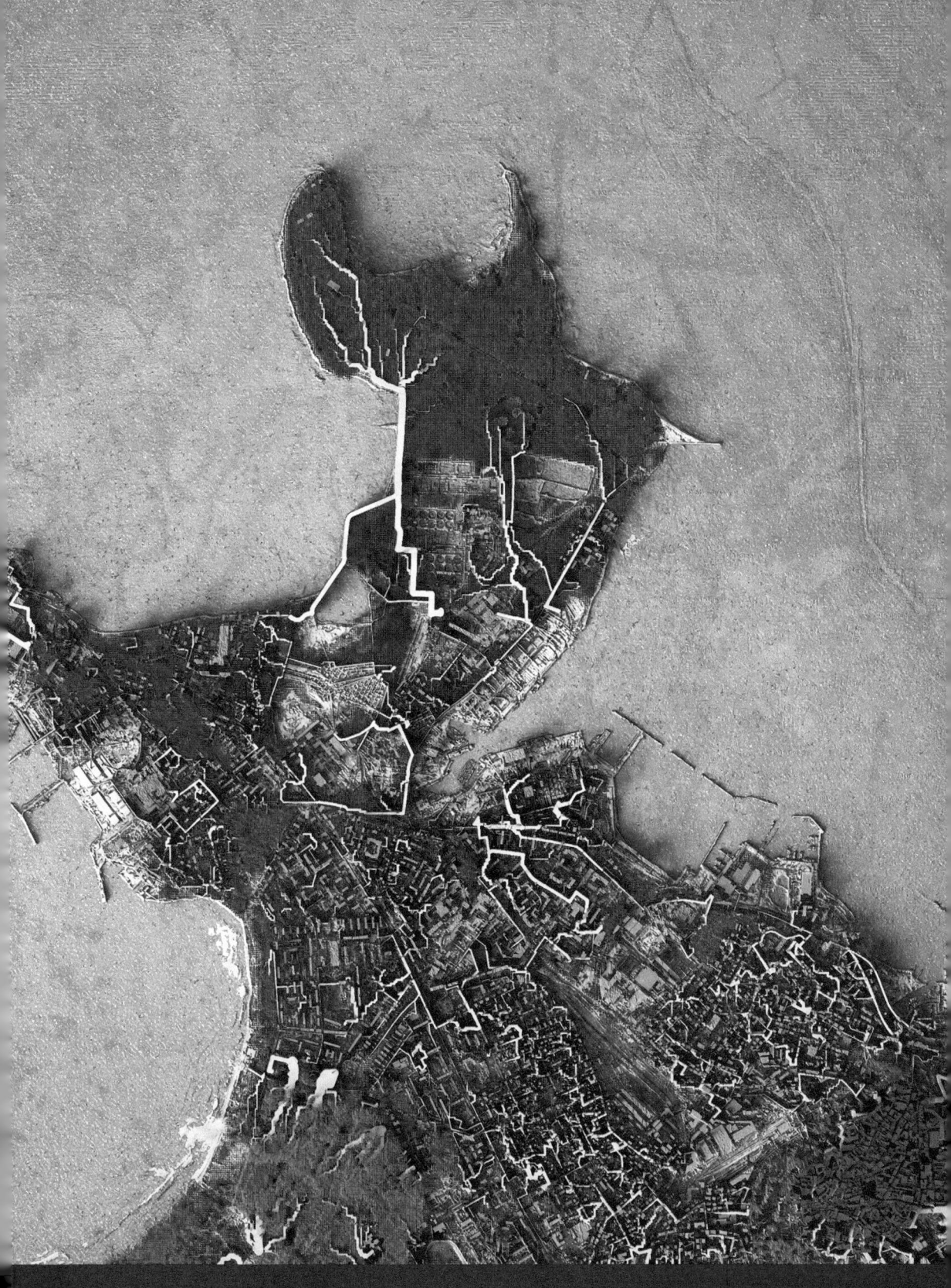

Aerial view of Tallinn Wet City. The new urban morphology proposes bio-architectural links envisioned as bridges across a flooded urban landscape during the summer months of 2050.

Ground plan of the Anthropocene Island proposal for the Paljassaare Peninsula, Tallinn. Colours index different stages of the wastewater treatment and pathogen density. The green colour range provides an indication of its cleanliness from wastewater pathogens and its fertility.

The 'Ground' protocol for Paljassaare Peninsula proposes the morphological hyper-articulation of the existing landscape and its living systems. Constantly monitored via satellite, this synthetic urban landscape feeds back to Tallinn's wastewater network in real time. Each molecular transaction has its spatial location, morphological effect, informational address and ecosystemic value. The process starts with the ESA (European Space Agency) supplying Level 1 data from the satellite Sentinel-2 at resolutions of 10x10 metres for an area of 3x3km thus framing the peninsula. Each pixel represents a degree of biochemical activity defined as 'wetness' and computed with the Normalised Difference Water Index algorithm. The resulting gradient field is indexed at specific locations along its ISO lines at a resolution of 2m. In each location tendency lines are computed: the longest lines appear in the areas of highest difference in wetness or biological activity. Such locations possess maximum potential for thickening and articulating into biochemical reactors.

These prototypical bundles for wastewater purification and sludge bio-digestion are equipped with active biotechnological units. The system is monitored in real time sending information about the status of its internal metabolism and receiving updates from the wastewater treatment network. Its operations are constantly altered and adjusted by distributed sensing /digging robots, the 'cyber-worms'. The articulation of the existing landscape determines directions of flow and purification; and in areas where the concentration of active bio-digestors is high and their emissions of heat and nutritious soil reaches a critical mass, new microclimates and related habitats are formed. Growing plants, insects and birds are attracted and become active agents of urban transformation.

The 'Air' protocol is ready to begin. Swarms of stigmergic building drones respond to these atmospheric gradients and begin the erection of new superstructures by lifting into place modular units of different size and form, each of them hosting an inhabitable bio-reactor cell. These flight-assembled ground-scrapers can evolve into assemblages of many thousands of units, symbiotic to their counterpart in present day Tallinn. They will be fed by the city's wastewater and in turn feedback natural gas, food and fertile soil. The urbanisation of Paljassaare becomes the biotechnological re-metabolisation of Tallinn itself.

Bird's eye view of the Anthropocene Island development on the Paljassaare Peninsula in summer 2050.

Bird's eye view of the Anthropocene Island development on the Paljassaare Peninsula in winter 2050.

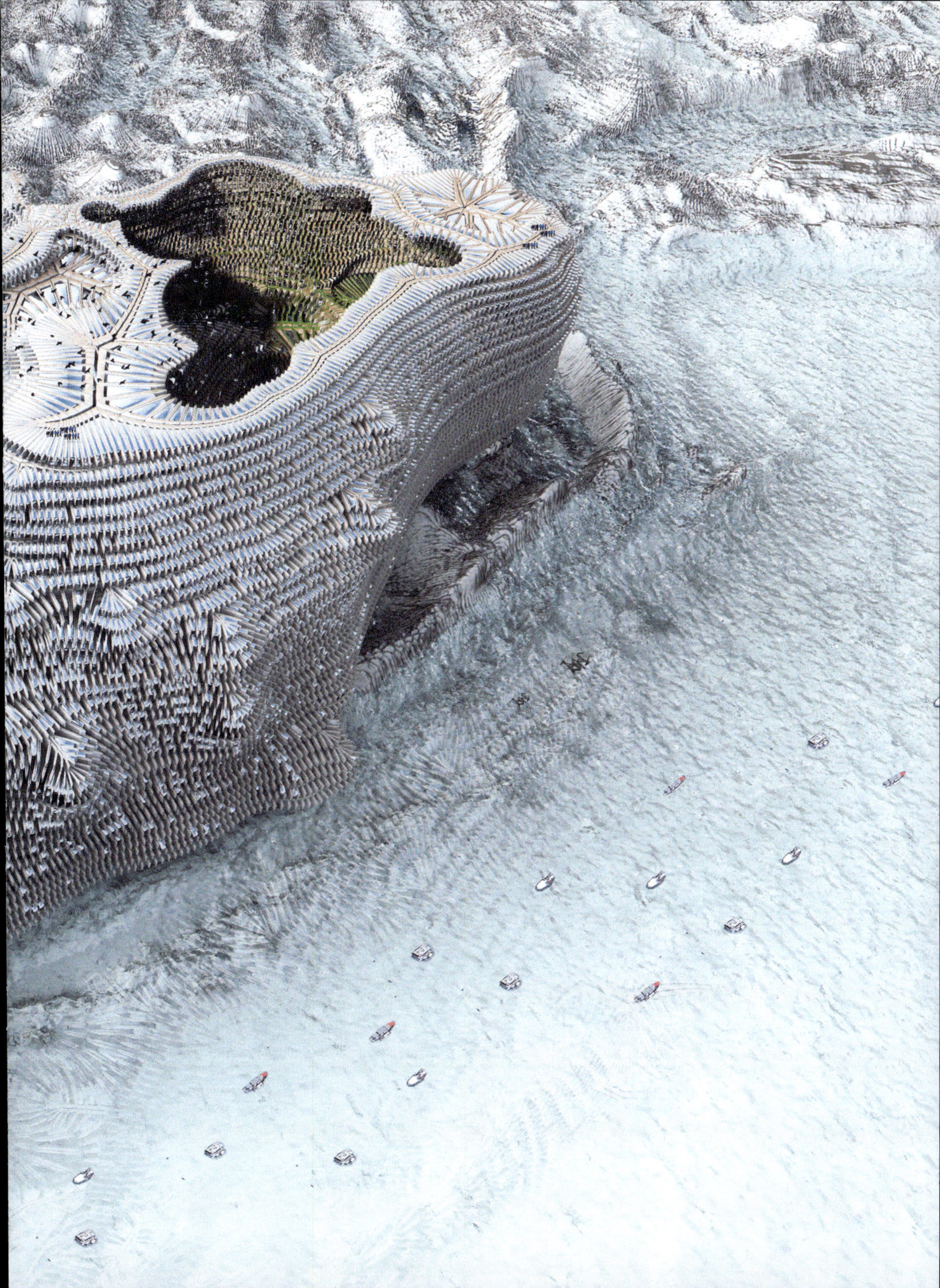

♪♪ Biocoenosis Nest is an association of different organisms in a non-invasive integrated community that establishes controlled coexistence in a post-natural scenario.

2.01.04

Biocoenosis Nest: A city woven by collective intelligence

Oscar Villarreal

For the past 200,000 years, humans have gained control over other planetary ecosystems and have become the most impactful species on the planet. This evolution initially brought us big benefits. However, it is currently generating situations that threaten the survival of many living entities including humans. Since humankind started to invade wild habitats, the boundaries between ecosystems became blurred. This form of invasive coexistence is having side effects such as the recent COVID-19 global pandemic.

In the past, urban design theorists have exemplified the ways in which urban design is approached by human intelligence with the purpose of making the human experience better without adequately considering other entities and their environment. Now, however, a non-human-centred approach is necessary in order to consider other forms of intelligence and create planning strategies beyond human limitations.

London's East End, having been an overpopulated and polluted industrial area, has been chosen as the stage for a design scenario that explores the consequences of a possible catastrophic urban future for London. Through the exploration of probable climate change scenarios that could lead to flooding in the area, the 'Biocoenosis Nest' project embraces water to turn the local marshland into a node of biodiversity. This node restructures the relationship between water and soil in order to support habitat restoration.

The main aim of the project is to connect the marshland's housing developments to the inner part of the city. This was achieved by computing optimal habitat locations, thus materialising a proximity diagram connecting biodiversity nodes and existing green areas. The use of computational design strategies provides for an approach toward non-human-centred distribution of these habitats, and thus a more robust urban diagram emerges.

In this proposal, behavioural strategies inspired by slime mould behaviour were adopted to define the routing network that shapes the proposed terrain of the marshland city. The resulting network actualises in an eroded terrain, the post-natural landscape of the marshland city.

In order to design its complex coexistence of systems, the bird nest biological model is analysed and deployed to rethink human and non-human interaction. Its fibrous logic generates a new synthetic nest, forming the porous architectural volumes that host the multi-species nodes. These volumes are formed by combining the different programmatic typologies into a new fibrous architectural system.

This design approach helps us to understand factors that can otherwise be incomputable by rational human intelligence and enables the development of post-natural urbanism, a practice that prioritises the health of every living entity in the Urbansphere.

As such, the architecture of Biocoenosis Nest is woven by collective intelligence, thus responding to the needs and requirements of multiple living entities. It is an association of different organisms in a non-invasive integrated community that establishes controlled coexistence in a post-natural scenario.

Cover page: Biocoenosis Nest masterplan envisioning a new morphological development of marshlands in the East of London. The plan is simulated with the support of AI algorithms trained with datasets of slime mould biological networks.

This page: Vision of Biocoenosis Nest architecture is engineered as a composite fibrous structure. The proposal is constructed in bio-cement components that contribute to the propagation and growth of wildlife in the marshland landscape

🎵 The aim of the GAN.OS project is to explore the possibility of an alternative distribution of settlements while avoiding the conceptual trap of a simplistic distinction between the human and non-human.

2.01.05

GAN.OS 101

Emmanouil Zaroukas

GAN.OS 101 is a project that embodies with unique conceptual clarity the research work that is taking place at the Urban Morphogenesis Lab, one of UCL's leading postgraduate research clusters.

GAN.OS 101 stands for the Geological Adversarial Network of Settlements and is a highly speculative aggregation of different intelligences and their capacity to construct adversarial worldviews within and beyond the human. The aim of the project is not only to revise the borders of established and sovereign disciplines, but also to experiment with the possibility of new spatial distribution of a yet to come sociality. The GAN.OS revolves around three major theoretical issues of bio-design conceived by the UMlab research team: intelligence, inclusion, and the making of worlds.

"The oldest and strongest kind of fear is the fear of the unknown", fiction writer H.P. Lovecraft once wrote. The limits of human knowledge create distress and uncertainty. Our intelligence has evolved a navigational mechanism that not only relies on the simple collection of facts, it also renders those facts with hues that emerge from concerns and overall directions of humanity at any moment of factual extraction. Every fact is rendered by a specific concern. Human intelligence is shaped in this tension and this tension reveals that our intelligence is neither unique nor general. It is not an exclusively human privilege.

Cover page: Architectural vision of GAN.OS 101 in Kamchatka, Russia. The proposal explores the possibility for an inhabitable topography, a thick ground, rising above sea level. Its architecture is capable of dealing with the ebb and flow of the context, showing how the system can exist in various conditions within a volcanic active environment.

This page: Axonometric view of an architectural scale prototypical part of GAN.OS 101.
The structure is designed to host living algae colonies operating as a source of biomass and biofuel powering GAN.OS 101.

"We are indeed surrounded by intelligences that the human is unable to sense, perceive and grasp. The unknown becomes terrifying and therefore excluded. The limits of the human sensoria as it confronts the cosmos creates a world, a world that is not the world-for-us, the known known world, nor is it the Earth, the world-in-itself, the known unknown territory, nor is it the Planet without name, that is the world-without-us, the unknown unknown." (Thacker, 2013)

This is precisely the move that the project GAN.OS undertakes: it plunges itself into the realm of the unknown unknown, the hostile territory of volcanic activity becoming the field of 'roars'. Some are captured by our media while others are simply speculated about. What cannot be recognised falls in the category of the non-human. The aim of the GAN.OS project is to explore the possibility of an alternative distribution of settlements while avoiding the conceptual trap of a simplistic distinction between human and non-human. It starts without names, without pre-existing categories and concepts. It refuses to deal with the Gaia, the planet Earth, the Earthlings and the non-humans, and it attempts to move beyond the conception of ecology within the human enclaved economy of concepts that decides and effectuates the human and non-human.

The design proposal therefore refrains from concepts that for centuries have organised the production of knowledge in urban design, both in theory and in practice. It undertakes the difficult journey that starts without conceptual distinctions to eventually reach new ones. It is for that matter that immerses itself into the planet-without-name in order to observe multitudes of production of knowledge and to revise known conceptual apparatuses.

Sectional view of the prototypical part of GAN.OS 101 devoted to the production of biofuel and food. This underground strata section is made of volcanic clay substratum that propagates the growth of an algae- and mycelium-based energy and food production system. Cover Page, previous page and this page, images by Shusheng Huang, Tao Chen, Anshika Tajpuriya, Sheng Cao, Meng Zheng.

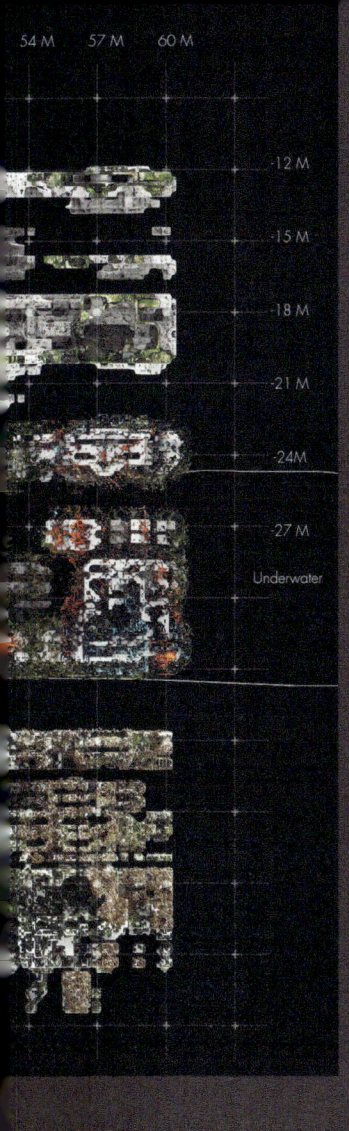

GAN.OS 101 engages the concept of coexistence and or 'commonality without inclusion'. Inclusion is recognised as a futile project that goes against the perpetual partiality of humanity's perceptual and epistemological capacity to grasp part of the planet and to turn that consequently into worlds. As such, the project refrains from mobilising a simplistic understanding of inclusion where humans and non-humans coexist. The project seeks a cohabitation of incompatible organic processes that goes beyond the mere inclusion of the non-human in the sheltering process.

The aim is not to simplistically and uncritically bring within the urban environment the non-human dimension, rather it is to invent new constellations of concepts that may spring novel spatial typologies. In that sense, GAN.OS 101 is developed on "a common task" (Wark, 2021) rather than the inclusion of others within a sovereign reality. Knowledge production is not the domain of a sovereign interiority of a specific world. "The common task is to produce knowledge of the world made up of the differences between ways of knowing it" (Wark 2021, p.4).

GAN.OS 101's take on bio-design is radical in that it enables an alternative approach to the incorporation of biotechnologies for the purpose of distributing constellations of new concepts which, in turn, redefine the human and the concept of design itself. GAN.OS 101 is the crystal ball rather than the mirror to our world, a refractory device in our making of words.

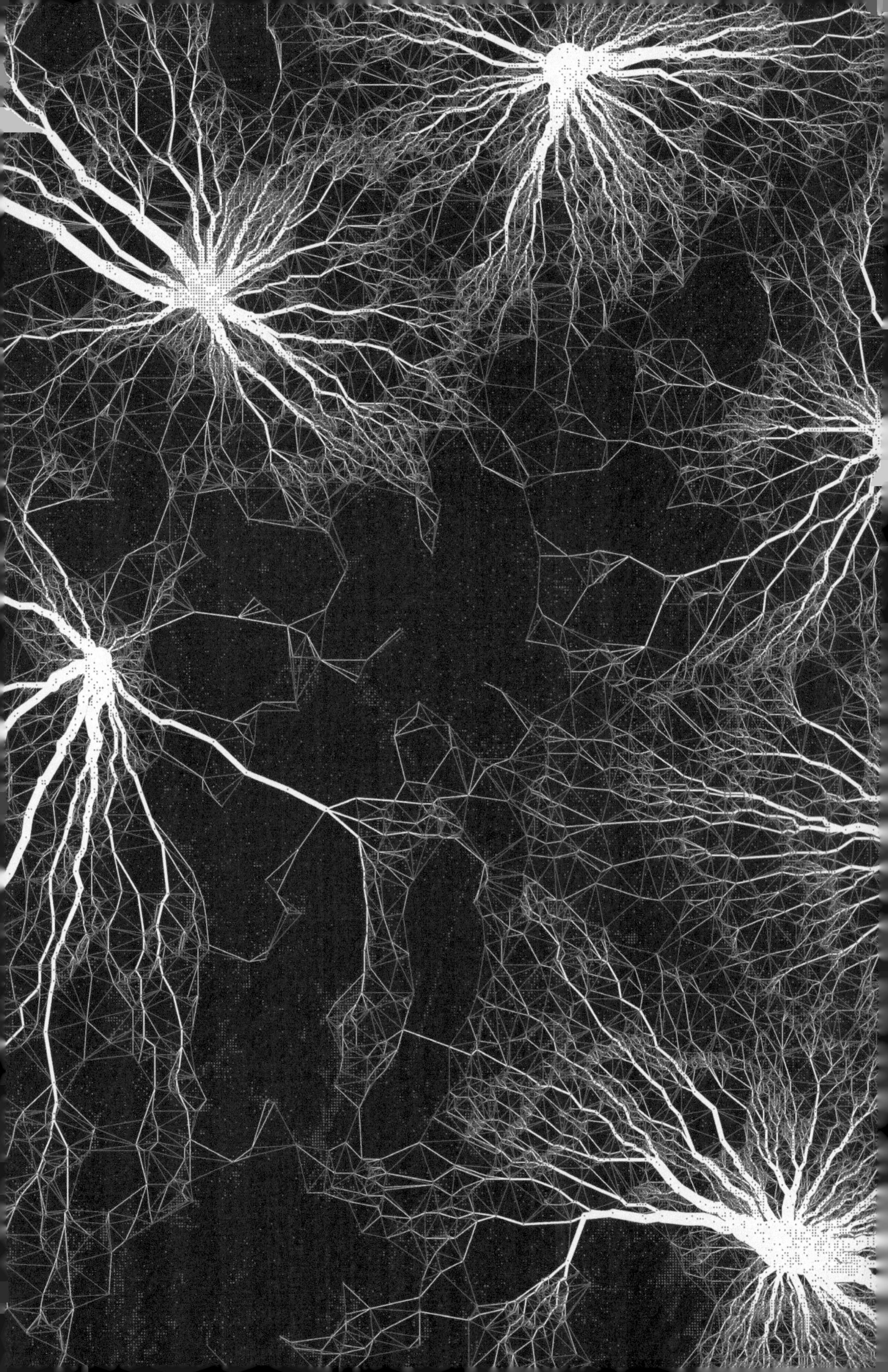

Projects 2.02

DEEP PLANNING

2.02.01 GAN-Physarum
La dérive numérique 306

2.02.02 DeepGreen
The case of Guatemala City 322

⚡ How can a slime mould shape the future of the Urbansphere?

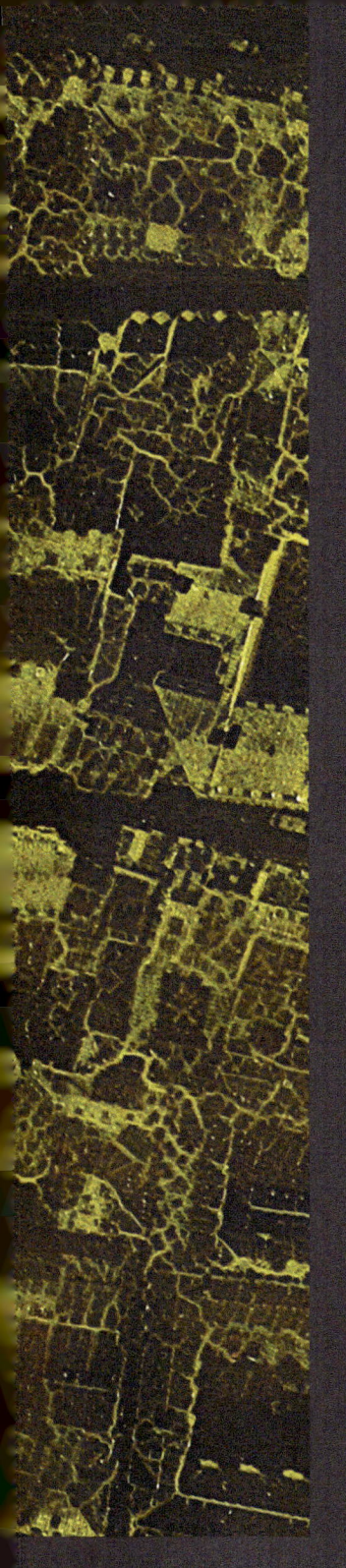

2.02.01

GAN-Physarum: la dérive numérique

GAN-Physarum: la dérive numérique, was unveiled within the exhibition 'Réseaux-Mondes' (Worlds of Networks) held at the Centre Pompidou in Paris (France) in spring 2022 and is now part of the museum's permanent collection. It depicts a future bio-digital, autonomous Paris. The vision, developed in conversation with curators Marie-Ange Brayer and Olivier Zeitoun for their series 'Mutations / Créations', questions the place of the network in our societies, its pervasiveness as well as its dematerialisation.

The network, it is argued here, is at the heart of technological change and several societal issues: surveillance, atomisation of the individual, actor-network, artificial intelligence and the emergence of the global Urbansphere. Against this backdrop, we are witnessing the unfolding of a post-natural history, a time when the impact of artificial systems on the natural Biosphere is indeed global, but their agency is no longer entirely human. Cities like Paris have become co-evolving networks of biological and digital intelligence, semi-autonomous synthetic organisms.

Technically a GAN (generative adversarial network) is an algorithmic architecture that creates new generative models using deep learning methods. This powerful form of artificial intelligence has been trained to behave like a *Physarum polycephalum*, a single-celled slime mould. When the trained GAN-Physarum is sent on a computational dérive on the streets of Paris, it shows us how to decode and reinterpret the gridded patterns of contemporary Paris into a smooth urban landscape. We witness a transition from the original morphological order to an emergent distributed network of evolving path systems.

Cover page: Satellite view of bio-digital retrofitting of the Centre Pompidou as imagined by the GAN-Physarum algorithm.

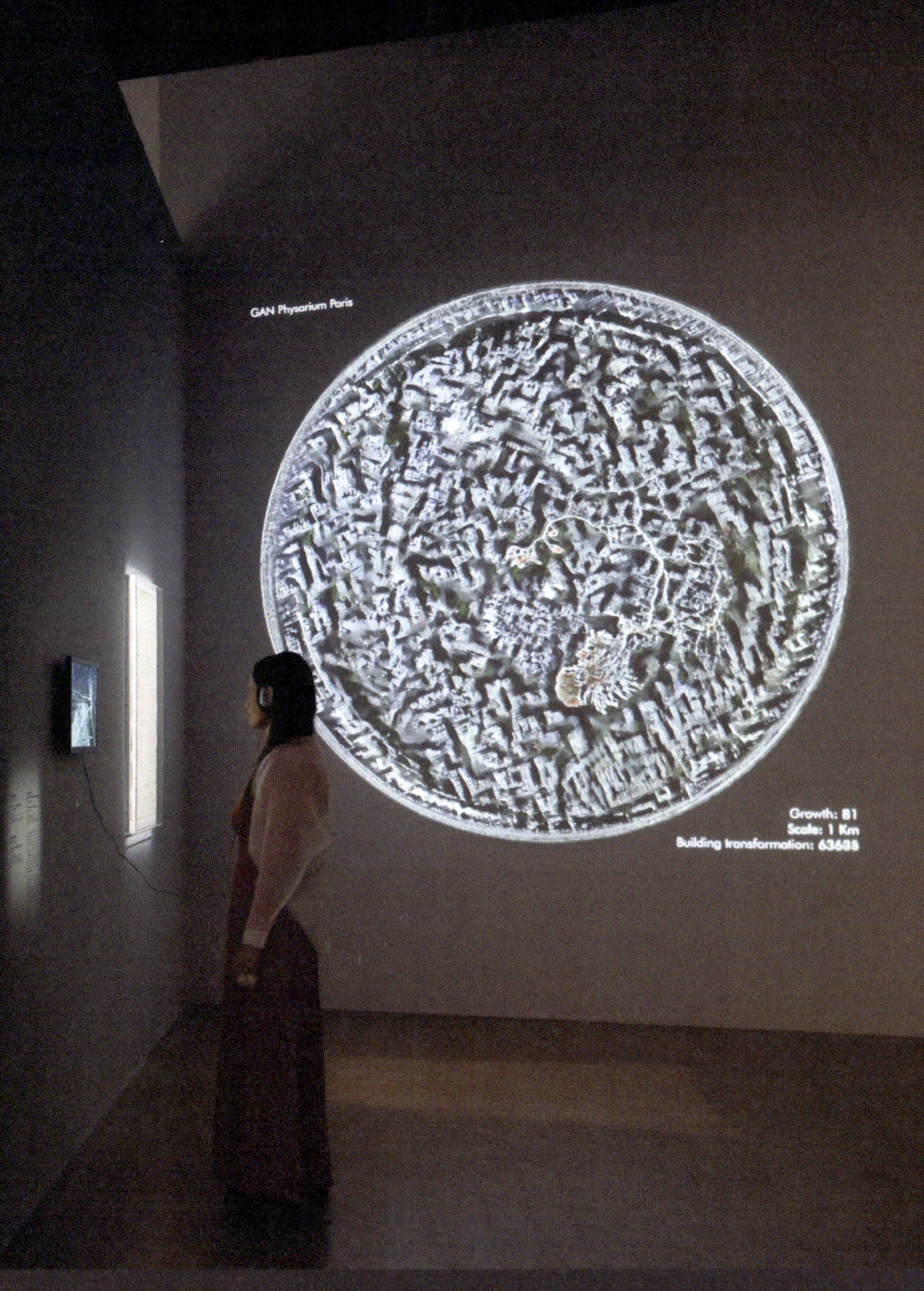

This spread: GAN-Physarum: la dérive numérique exhibited in the Worlds of Networks exhibition at the Centre Pompidou, Paris in February 2022.

GAN-Physarum: la dérive numérique, bio-painting, at Centre Pompidou, Paris.

Detail of the bio-painting highlighting a living *Physarum polycephalum* stretching its networked body to feed on a grid of nutrients, distributed on the canvas to accurately map current biotic resources in Paris.

The sequence, captured and brought to life in a computational time-lapse video, is accompanied by the corresponding bio-painting, approximately one square metre in size, where the living *Physarum polycephalum* stretches its networked body to feed on a grid of nutrients, distributed on the canvas to accurately map Paris's current biotic resources. The traces we can admire on the canvas at such high resolution are the embodiment of the slime mould's cognitive system, and a non-human narrative of a completely different urban structure, Paris's very own evolving biotechnological brain. It is perhaps impossible to capture this vison with words or a single image and therefore the value of GAN-Physarum is to provide us with an instrument to model it.

In Constant Nieuwenhuys words:

> "Thinking about a social structure that is so different from the existing one that it can safely be called its antithesis, words and terms are inadequate tools. Since what we are considering here is no abstraction but the material world, as in physics, it seems most logical to resort to visual tools, in other words a model." (Wigely, 1998)

In Constant's case this model was called 'New Babylon' and provides an extraordinary tale of a global city devoted to playfulness and creativity. Conceived in the 1960s as an original vision of a new way of life, the fictitious city was the product of a hyper-connected society that had achieved freedom thanks to its extraordinary apparatus of mechanical urban devices. While the machines handled all hard labour and productive work, New Babylon's population could indulge in an entirely ludic lifestyle.

This *Homo ludens* was a nomadic and free person. In New Babylon, this freedom allows for continuous movement and the possibility of living in a season-less, timeless, boundless, limitless environment. This aspect of nomadic life, according to Deleuze and Guattari, allows the inhabitation of a homogeneous smooth space as opposed to the organised striated space of the gridded city (Deleuze and Guattari, 2004). Moreover, nomadic space is haptic, associative, interconnected; just like the world the slime mould navigates.

Close-up of a living *Physarum polycephalum* in a petri dish. The organism is deployed as a biological computer to train an AI algorithm over a period of time.

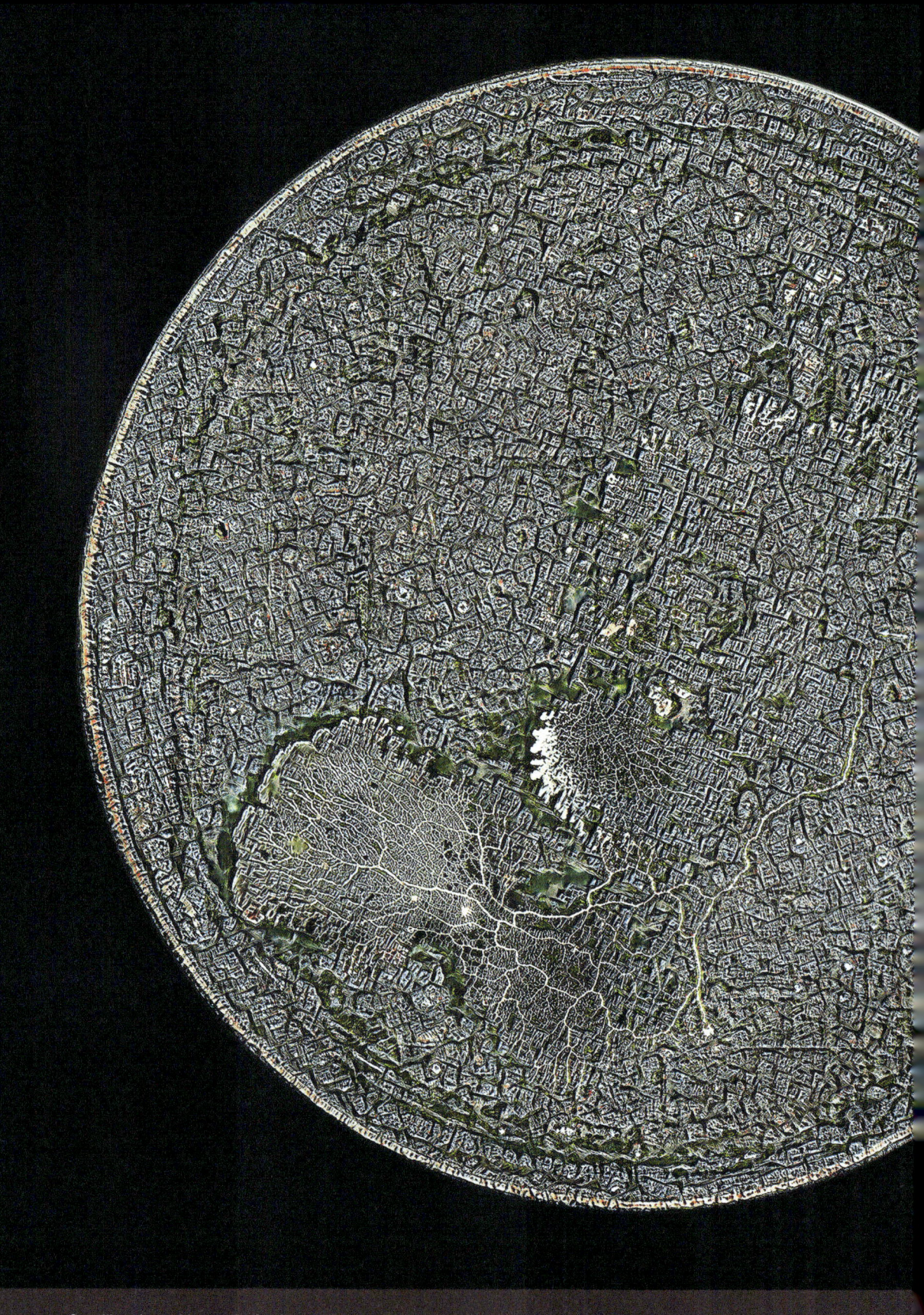

Satellite view of future Paris planned by the GAN-Physarum algorithm at a scale of 10km.

The slime mould is a mono-cellular organism with a peculiar body composed of thousands of nuclei afloat in a sea of actin enclosed by a single stretchy membrane. From this, it derives its unique morphing properties. Its first dérive took place in the air-conditioned laboratories of the Synthetic Landscape Lab at the University of Innsbruck. Here at a constant room temperature of 20 degrees centigrade a living *Physarum polycephalum* in its active plasmodium phase hunted for nutrients within the sterile environment of a borosilicate glass petri dish.

The process starts with a searching phase during which the pulsating body branches out in all directions detecting food sources and their relative distribution and size. What follows is a phase of optimisation. Finely detailed branches emerge in the relevant areas of the petri dish while other parts are abandoned. Eventually, some of the branches grow in size and become thickening convoluted transportation arteries. However, the optimised configuration never settles. As the resources diminish and their overall distribution changes, *Physarum polycephalum*'s morphology adjusts in real time.

The scarcer the resources become the more accelerated the change and adaptation. At the tipping point, *Physarum polycephalum* is seen racing around the petri dish in an attempt to find new sources with sufficient energy to sustain its searching effort. Once nothing is left, it retreats devoting all remaining energies to create new fruiting bodies, thus commencing the next phase in *Physarum polycephalum*'s uniquely elaborate life cycle.

On a philosophical level GAN-Physarum is an investigation into Guy Debord's theory of *la dérive* (Debord, 1958) and the effects of non-human psycho-geography on the perception of the city. Then, as now, performing dérive experiments and documenting the trajectories of the routes allowed for a new interpretation of zones as vortex, areas of differential tension present in the liquefied space of the city. In the slime mould, the spatial vortices have the ability to contract and expand, which is a result of the peristaltic flow within the liquid space of its body.

In GAN-Physarum's experiment, the main Parisian dérive started when a satellite image of Paris, duly processed to extract information about its biotic layer (plants, grasses, rivers and other 'wet' surfaces), was remapped on a physical grid providing the exact nutrients' distribution. The points on the grid transfer the geo-data of corresponding latitude and longitude of the map of Paris onto a canvas. A 3D data matrix is compiled storing the density of biomass as a percentage value of a fully vegetated pixel (one pixel is equivalent to a 10x10m area of Paris). These percentages of density are translated proportionally into amounts of nutrients on the growth canvas of *Physarum polycephalum*, with a cell size of 3cm. *Physarum polycephalum* was then introduced, beginning its long bio-computational process.

Similarly, in New Babylon, the nomad moves into an artificial, wholly constructed environment, where social mobility builds a kaleidoscope of events. One navigates the boundless city of New Babylon using psycho-geography, the practice of a subjective apprehension of space. Whether consciously or not, the environment is registered and re-organised from a series of geographical and physical fragments, individual emotions and affects. Constant's New Babylon was his proposal for the Situationists' response to the city in which traditional architecture has disintegrated and has been replaced with a vast network of multi-layered spaces which would eventually cover the whole planet, and in an endless Urbansphere (Nieuwenhuys, 1974). In GAN-Physarum, the authors deploy generative adversarial networks (GAN) to model an autonomous bio-digital city, a city made for both human and non-human citizens and planned by a new form of non-human artificial intelligence.

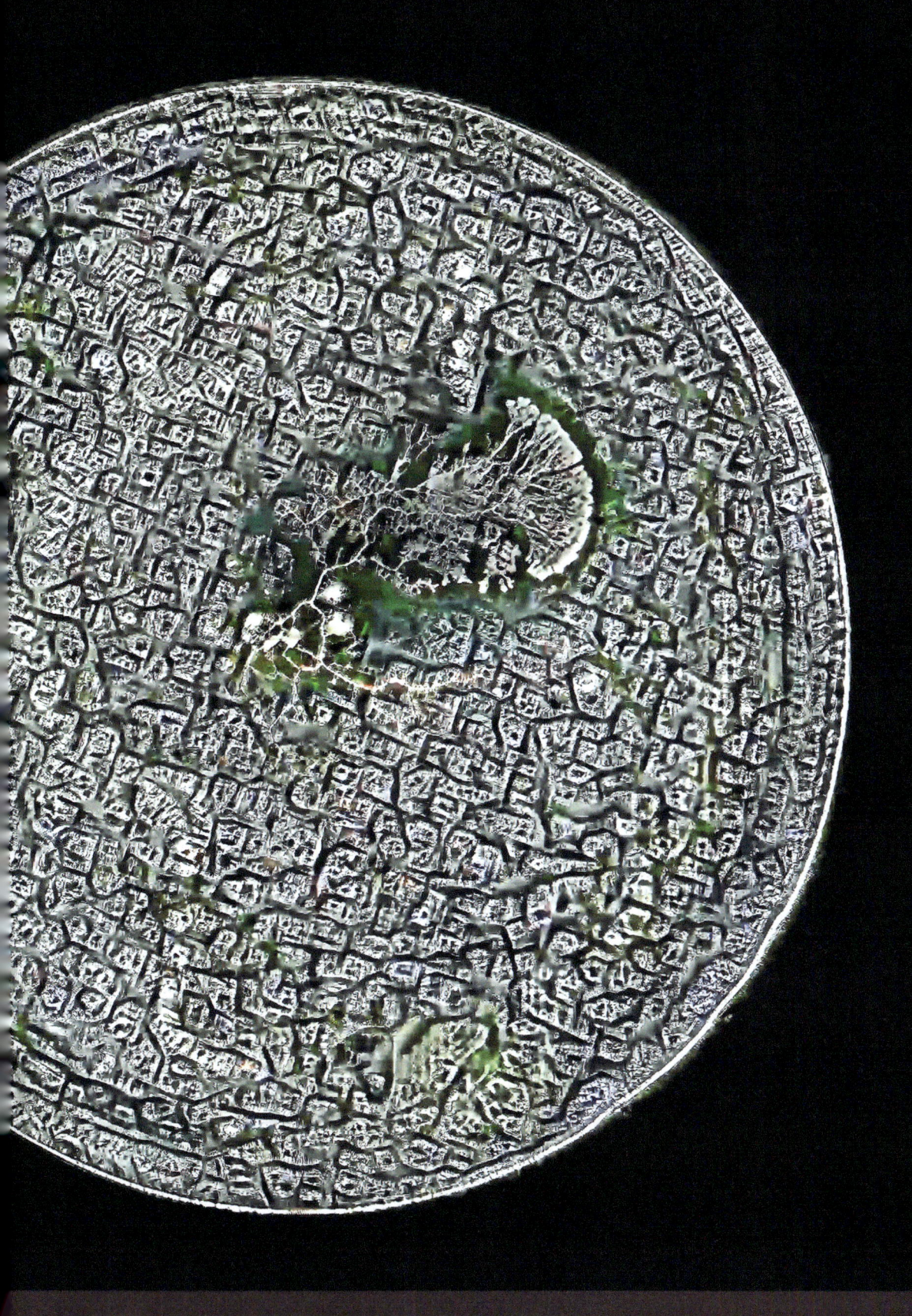

Satellite view of future Paris planned by the GAN-Physarum algorithm applied at the 1km scale of Paris. The simulation highlights the organisation of individual building blocks as part of a new bio-digital urban infrastructure.

Satellite view of future Paris planned by the GAN-Physarum algorithm applied at a 100m scale. The simulation highlights a complete redesign of the Centre Pompidou and surrounding public space as a bio-digital architecture.

GAN-Physarum deploys a machine-learning technique that uses the training of image to image translation models without paired examples. A GAN has two parts, the generator and the discriminator, engaged in an internal competition. The generator learns to generate plausible data. The generated instances become negative training examples for the discriminator. The discriminator learns to distinguish the generator's fake data from real data. The discriminator penalises the generator for producing implausible results (Radford, Metz and Chintala, 2016).

Depending on the training input, this process results in the transfer of key features from one set of images onto another, and vice versa. In the case of *la dérive numérique*, the domain of source images and target images refer to the slices of two actual input images that belong in two different domains, the urban and the biological. This technique transfers the behavioural patterns of the *Physarum polycephalum* onto the urban structure of Paris. The objective is to investigate how its biological intelligence can be applied on different scales to reinterpret existing infrastructures and building distributions of Paris itself, thus coupling what is biologically grown to what is algorithmically drawn.

While time-lapse imagery, as described above, captures *Physarum polycephalum* at different developmental stages, Paris is analysed at different resolutions through remote sensing. A multiplicity of urban structures and urban morphological patterns can be detected while zooming in to the Centre Pompidou from a wide frame that includes the entire centre of Paris. By using the power of ten as the decreasing degree of magnitude, the protocol captures frames of Paris at four different resolutions. The largest city scale is a 10 by 10km frame, including the entire circular Périphérique of Paris and centred on the site of the Centre Pompidou. While zooming in, further resolutions are registered, including a 1 by 1km frame, showing the Centre Pompidou within its surrounding neighbourhood, a 100 by 100m frame, representing the structure of the Centre Pompidou with the adjacent streets and squares, and finally, a 10 by 10m frame, focusing on the mechanical systems forming the external envelope of the celebrated Parisian architectural machine.

The machine-learning algorithm of GAN-Physarum is trained at each of these urban resolutions to a corresponding behavioural pattern of *Physarum polycephalum*. Over several epochs, the algorithm learns how to reinterpret *Physarum polycephalum*'s behaviours in relationship to Paris's morphological structures, and vice versa. This non-human cognitive process is underpinned by the GAN-Physarum's workflow developed by the authors and their team.

This workflow can be described in four main phases. First, the preparation of the input images for training purposes. The dataset preparation is automated, slicing input images from both time-lapse and satellite source domains into equal tiles of 256 x 256 pixels in size. The second phase is the training of the GAN-Physarum based on these input datasets. At this stage, the model is sufficient for generating plausible slices in the target domain. Subsequently the trained algorithm is tested. During this phase, the GAN-Physarum projects an input image in the biological domain A to an image in the urban domain B, and vice versa. Several cached models from the training phase can be loaded at this stage, thus tracking the self-development of GAN-Physarum. In the fourth and final phase the output slices are automatically recombined into true colour speculative satellite views.

While, at first sight, the visions conjured by GAN-Physarum have the disorienting quality of the non-human mind that conceived them, these connect us with contemporary Paris at a more fundamental level. To paraphrase Constant, disorientation is favoured here as a design practice as it furthers adventure, play and creative change. The space of GAN-Physarum's Paris has all the characteristics of a labyrinth, within which movement no longer submits to the constraints of the given spatial or temporal organisation and becomes the direct expression of social independence (Nieuwenhuys, 1974).

Satellite view of future Paris planned by the GAN-Physarum algorithm applied at a 3km scale. The simulation highlights the transformation of neighborhoods and new urban massing.

✿ DeepGreen implements the interdependence of digital and biological intelligence in urban design by pairing what is algorithmically drawn with what is biologically grown.

2.02.02

DeepGreen
The case of Guatemala City

In a recent collaboration with the United Nations Development Programme, ecoLogicStudio has been testing the potential application of artificial intelligence to develop a new planning interface named 'DeepGreen'. Its aim, derived from the implementation of the GAN-Physarum workflow introduced earlier, is to transform planning into a participatory process informed by a big data analysis of the trans-scalar nature of the contemporary urban landscape.

The DeepGreen protocol is designed to register multiple inputs so that a diverse range of stakeholders can interact with several layers of urban data and test their hypothesis across various scales. DeepGreen simulations are time based, enabling all stakeholders to appreciate the effects of new policies and strategies systemically. Finally, it is built onto a visually compelling and virtually immersive interface, deploying VR technology to invite all stakeholders to experience simulated urban scenarios.

These characteristics are critical to help promoting the evolution of new planning practices with restorative potential on existing urban systems. The demand for such restorative practices is increasing globally since more than half of the world's population now lives in cities and the urban population is expected to double by 2050. Intense urbanisation is forcing profound transformations in human socioeconomic developments on a global scale. Particular emphasis is now given to the relationship between human beings and the urban landscape that sustains their livelihood.

The result of this radical transformation is that cities and city regions today are at the forefront in experiencing the effects of global climate change. However, perhaps paradoxically, cities are also emerging as cradles of a new civilisation that seeks to establish a more sustainable relationship with the Biosphere. It is therefore necessary to redesign their infrastructure, rethink consumption patterns and implement circularity within what we have defined as the Urbansphere. This entails innovative strategies of waste management, water conservation and recycling, renewable energy production and trading.

From the fieldwork that the authors have conducted over the past twenty years in cities such as Caracas it has become obvious that the significant resilience demonstrated by fast-growing metropolitan regions in the face of extreme adversities is often due to the layers of informality which supplement and complement existing public services. Whether it is in water catchment, individual waste recycling, or in other forms like decentralised construction, these dynamics are an essential part of the tapestry of life in cities. Contemporary planning methods and policies however are often unable to recognise and capitalise on these efforts. Yet, effective ways of addressing vulnerabilities demand utilisation of all forms of intelligence, human and non-human, individual and collective.

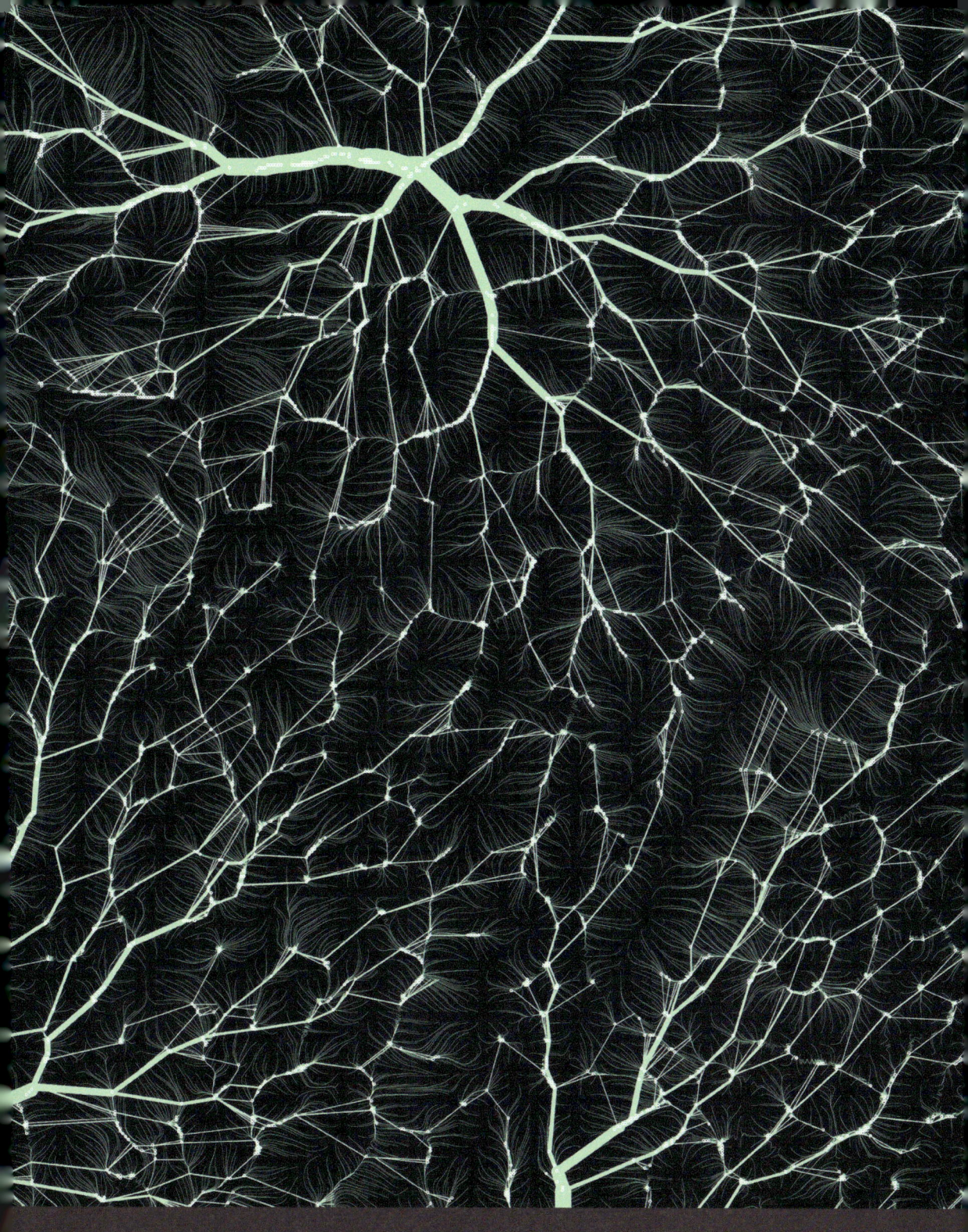

Cover page: Satellite view of the re-greening plan of Guatemala City. The vision is achieved through the algorithmic manipulation of the existing urban fabric with artificial intelligence.

This page: Guatemala City Blue-Green Masterplan. The new rainwater collection and purification infrastructure. The image is computed through a combination of water flow simulation patterns and minimal networks on the digital elevation model of Guatemala City topography.

Guatemala City Blue-Green Masterplan. The plan highlights the new local to municipal waste collection networks. The plan is algorithmically computed from GIS map, satellite map and digital elevation model analysis. The analysis also takes into account emerging local waste collection hotspots and is updated in real time.

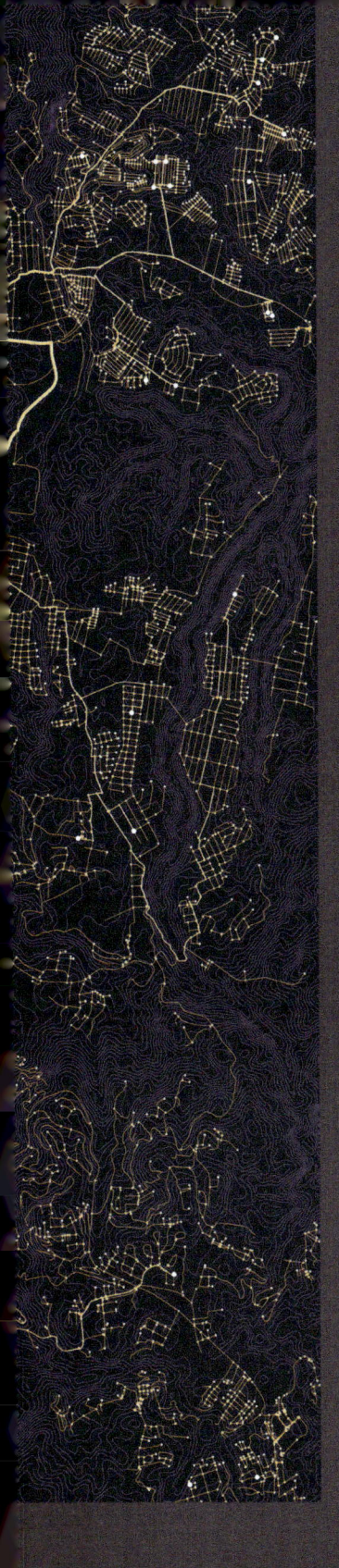

The DeepGreen protocols enable us to design resilient cities that use their size and collective energy to create refuge for both humans and displaced wildlife, that promote the emergence of positive microclimate, that replenish depleted water sources and that restore degraded terrains. Processes such as desertification, land erosion and contamination can be re-metabolised by innovative strategies of urban re-greening and re-wilding as well as of urban agriculture. Ultimately, DeepGreen provides a method to mobilise collective agency and intelligence in order to face the challenges ahead. To illustrate how DeepGreen pioneers a new application of sophisticated machine-learning algorithms to achieve these goals, the Guatemala City project provides a poignant case study.

Guatemala City is situated on a complex and highly unstable terrain surrounded by mountains and volcanoes, some of which are still active. Its ecosystems, originally very rich in biodiversity, are now made fragile by unchecked urbanisation and, given its climatic zone, the effects of climate change. In Guatemala City this scenario is exacerbated by a serious lack of waste management. The municipal garbage dump is the biggest landfill in Central America containing over a third of the total garbage in the country. 99% of Guatemala's 2,240 garbage sites have no environmental systems and are classified as 'illegal'.

In this specific case, the DeepGreen protocol is deployed to interface bottom-up processes of self-organisation, such as the many local waste-recycling activities that are emerging out of necessity in the areas closer to the dumping sites, with the strategic decision-making that occurs at municipal, national and international level. The aim is to find new synergies and direct investments where and when they have the most potential to engender positive change.

To begin with, the project indexes different kinds of remote sensing data: a land survey (at a resolution of 10m); digital elevation models (at a resolution of 30m from the NASA database); and GIS vectorial data from the Open Street Map project. These datasets are processed at a resolution of 1m over a frame of 16km comprising the entire city centre and several peripheral neighbourhoods. Municipal, local and informal waste disposal and recycling sites are mapped and their location marked. The existing road network is also mapped and all potential sources of waste are located.

Two sets of path systems are then computed: local path systems connect all waste sources with the closest dumping sites. Municipal path systems connect all dumping sites with six proposed municipal recycling centres. All paths are thickened in a colour-coded graphic representation to highlight the amount of waste that travels on each of their branches. The process is repeated for different kinds of waste—organic, metal, paper and plastic. The emerging diagrams, algorithmically updating in real time, are fed into GAN-Physarum as urban datasets.

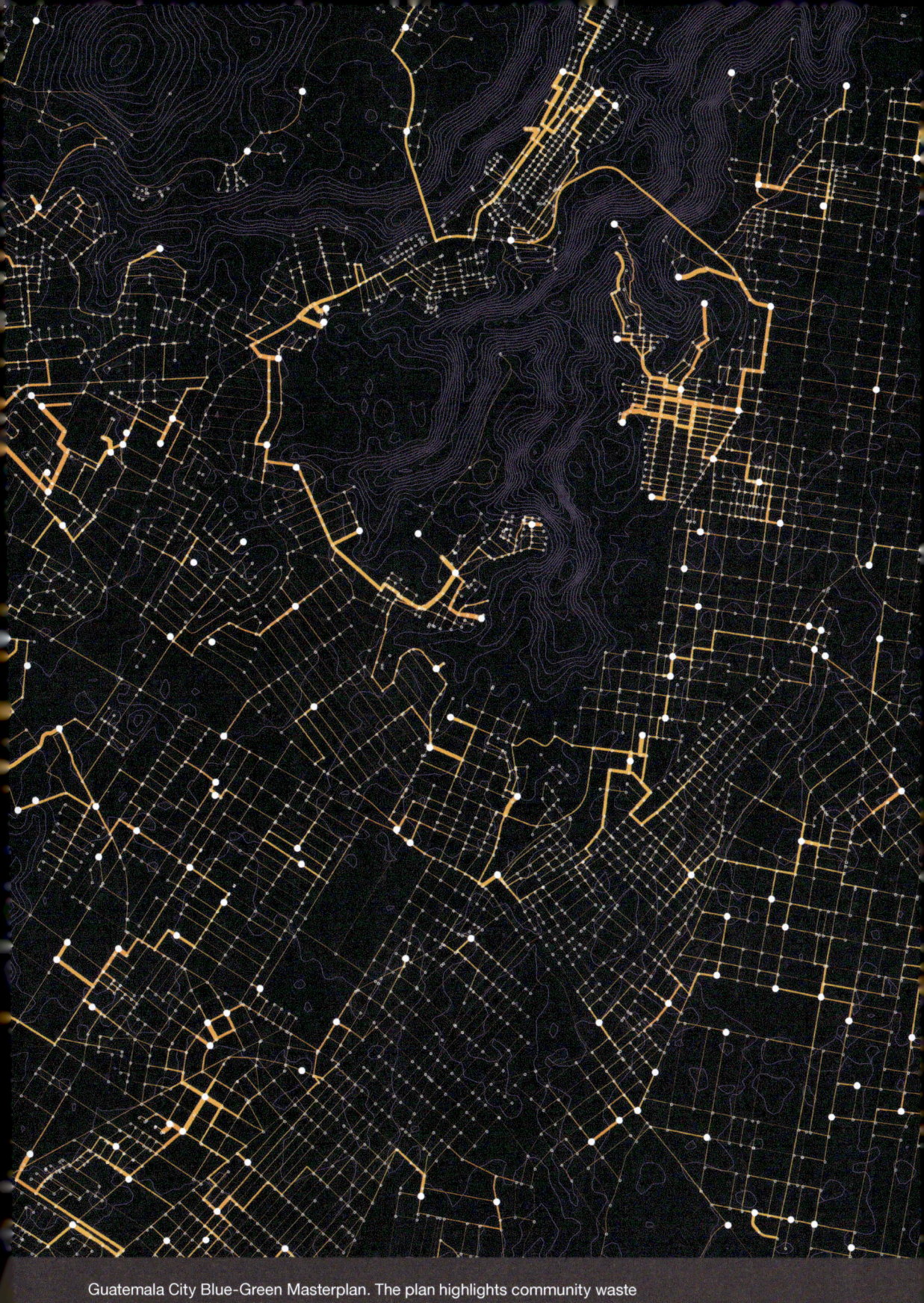

Guatemala City Blue-Green Masterplan. The plan highlights community waste collection networks and is updated in real time through algorithmic simulation.

Guatemala City Blue-Green Masterplan. Biological waste infrastructure. The municipal waste collection networks of Guatemala City are visualised as a living organic infrastructure that co-evolves with the natural landscape. The vision is developed using the GAN-Physarum algorithm.

Guatemala City Blue-Green Master Plan. Biological waste infrastructure. A new urban morphology is obtained by co-evolution of two overlapping systems: the local to municipal waste collection networks and the vegetation network of Guatemala City.

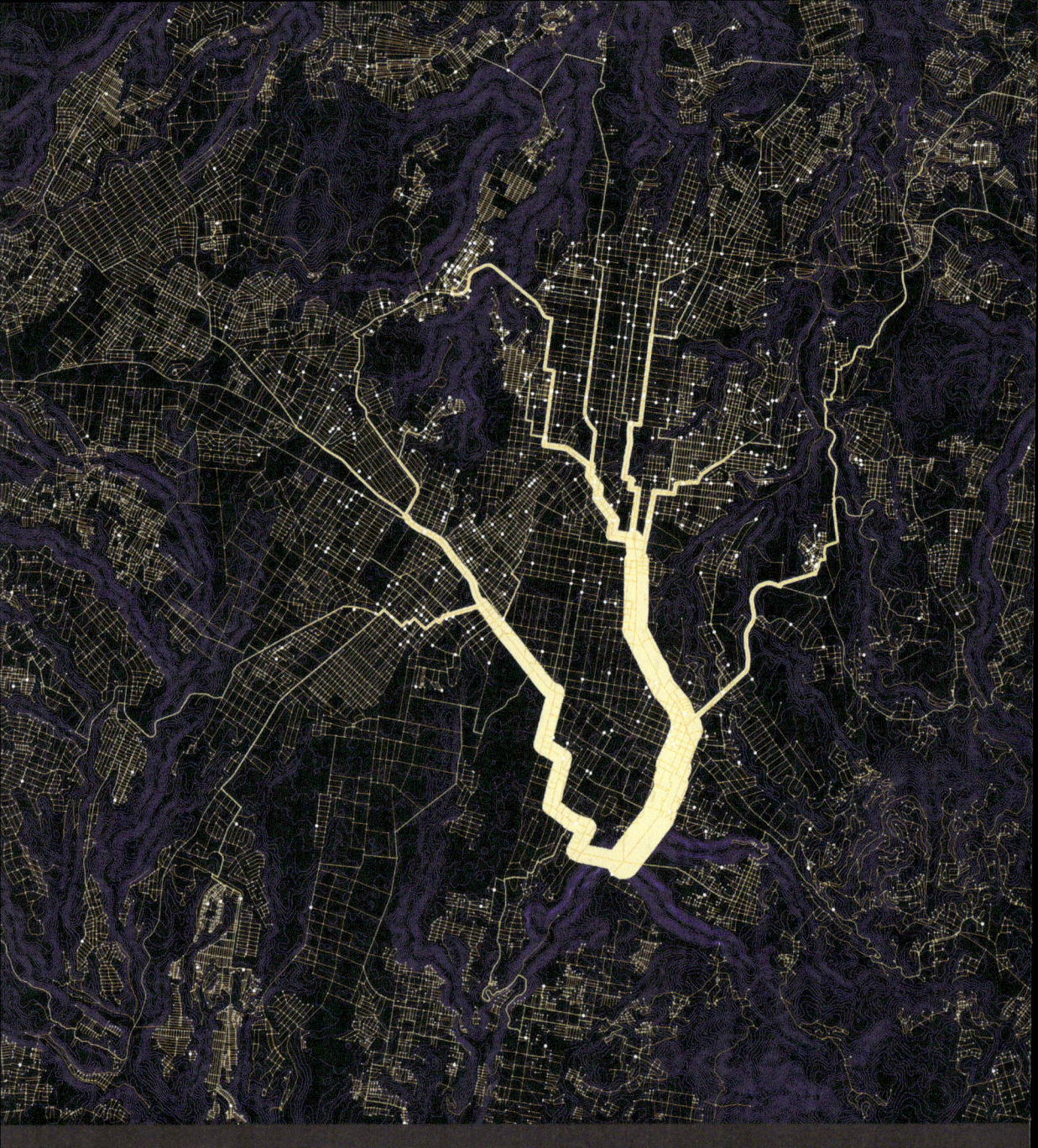

Guatemala City Blue-Green Masterplan. The plan highlights the new local to municipal waste collection networks.

Guatemala City Blue-Green Masterplan. The new biotic network. The green plan is computed with a proximity algorithm from the accurate mapping of existing biotic systems. It highlights areas lacking connectivity and requiring re-greening.

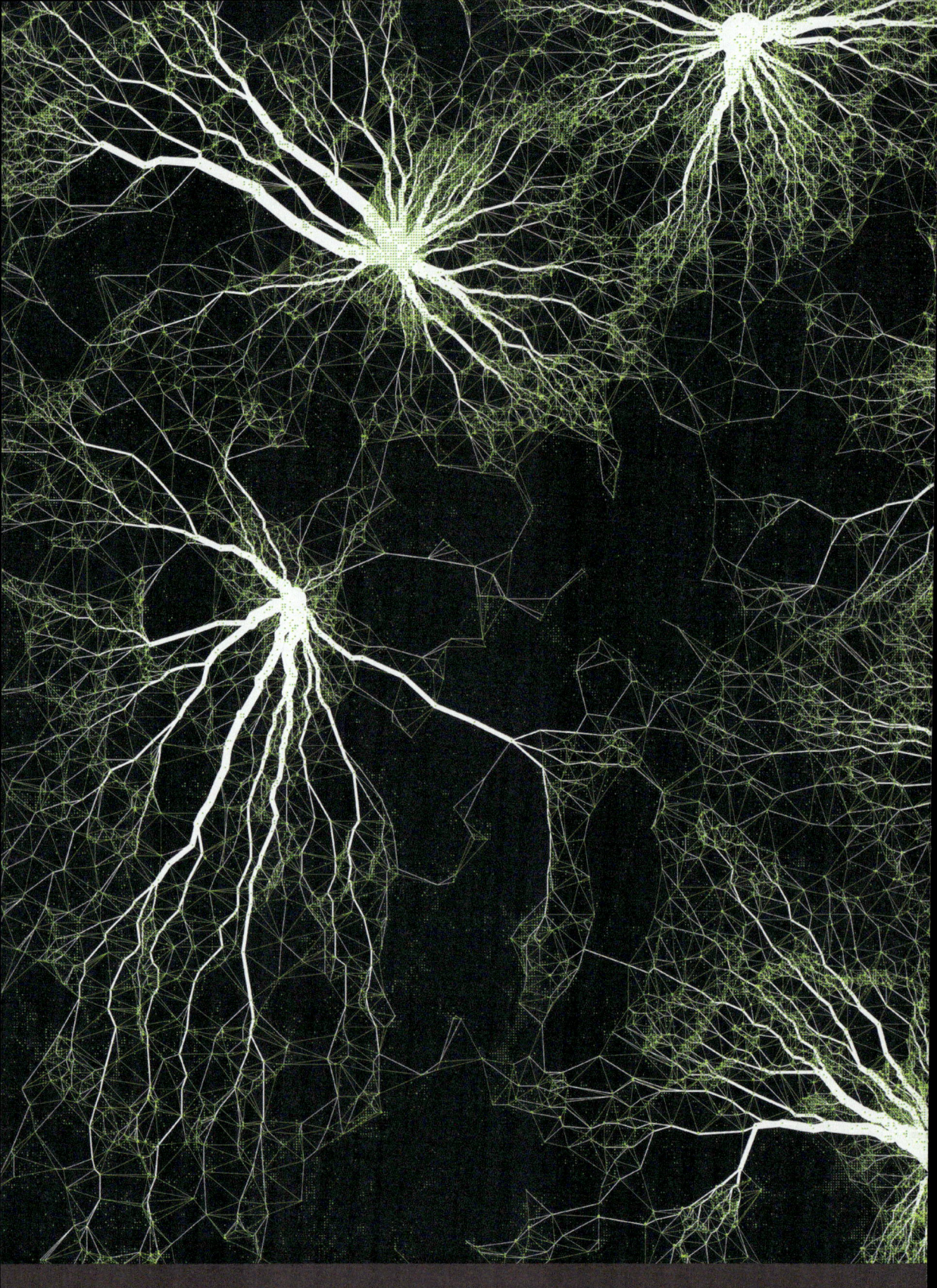

Guatemala City Blue-Green Masterplan. The new biotic network. The plan highlights the formation of new green corridors for urban agriculture and wildlife restoration. Thicker lines indicate denser biotic networks in the city.

After several iterations of the DeepGreen protocol, two radical visions emerge from the case of Guatemala City. On the one hand, the waste infrastructure appears as a biologically active, convoluted and highly differentiated body, capable of sorting, transporting and re-metabolising the urban waste into nutrients and raw materials. On the other, GAN-Physarum reimagines Guatemala City as a networked body gently suspended over a wild substratum of proliferating biological life. Once shared with a wider group of stakeholders these visions have triggered a more concrete set of proposals among which two are finding significant traction: a re-wilding plan to foster new coexistence between human and urban wild animals; and an urban agriculture plan that proposes a method to guarantee food security and recycle organic waste while employing the rural population currently migrating to the city of Guatemala.

The design process is ongoing, but what matters thus far is that both proposals are sensitive to local conditions while effecting international power relationships. For instance, migrating birds populating Guatemala City migrate to and from Canada. Therefore, investments in urban re-wilding will benefit the biodiversity of Canada too. Similarly, migrant workers cross the city on their way to the US–Mexico border. Urban agricultural plans could retain rural workers, thus alleviating the pressure on both Mexico and the USA. Such synergies have the potential to channel significant international funds to local projects, thereby improving the life of citizens of the Guatemala City region at large.

Satellite view of the proposed biological waste infrastructure at the neighborhood scale in Guatemala City.

Satellite view of the proposed biotic network at the neighborhood scale in Guatemala City.

Prof. Mario Carpo

Tomatoes and the immortality of the soul

There is an anecdote I often tell my students to explain how today's wave of computational automation may, if we are smart, lead to a new and better way of making. Imagine that you are a tomato farmer, and you want to choose today's best technology to grow and sell your tomatoes at the lowest cost. The traditional, modernist solution would read, ideally, as follows: first, you should genetically modify your tomatoes, so they would grow to standard, easily manipulable shapes and ripen at the same pace. An engineer will likely prove that the best tomato shape for picking, sorting, packaging, and shipping is cubic. You will then grow a crop of genetically modified cubic tomatoes, and hire an industrial robotic arm, driven by a scripted algorithm, to shake rows of identical cubic tomatoes, all in the same shape and at the same point of ripeness, cull them from their hydroponic vines, and push them onto a conveyor belt. Another industrial robot would then pick these cubic tomatoes for boxing, labelling, and shipping. A truck would convey these wares to a logistics hub, from where they would be dispatched to fulfilment centres around the world by road, air, or rail. This is what a modernist tomato grower would do. This is more or less what many tomato growers today do, regardless of ideological allegiances; this is, I suspect, what Rem Koolhaas would do, given the chance.

But today there is an alternative. Today we can use a more versatile, adaptive, intelligent technology to make the entire process fit the natural variations of a natural product. Let tomatoes be tomatoes. Let them grow as they always did. Then let's train an automated harvester, driven by robotic vision, sensors, and AI, to scan each tomato as my great-grandmother, I presume, would have done: picking each at the right point in time for putting in a basket and selling at the local market. My great-grandparents lived a few miles from town, so they commuted to the market square by carts and bicycle. Today, a fleet of electric self-driving vehicles could easily do the job; the local market, for all I know, is still there—albeit reduced to a caricature of what it used to be. Using computational automation, we could bring it all back to its original function: providing fresh local food for local residents.

The next frontier of robotic automation will not replace industrial workers repeating identical motions to execute the alienated notational work needed to produce standardised, one-size-fits-all, wasteful items of globalised mass production. That's already done; and we shall need less and less of that. Tomorrow's technology will beget a new kind of artisan worker, carrying out unscripted, endlessly variable, inventive, and creative tasks to produce no more and no less than the right amount of non-standard stuff we need: where we need it, when we need it, as we need it; made to specs, made on site, and made on demand.

Claudia Pasquero and Marco Poletto, while sharing my faith in post-industrial technologies, go one step further and venture into grounds where I may not always feel comfortable to tread. But this is the charm of their work, as they themselves aver its inherent ambivalence, and acknowledge the underlying tension between their apparent biophilia and a technical delivery aimed at quantifiable performance and market value. Yes, their work does at times betoken, in Sanford Kwinter's words, a sympathy for the "the rich indeterminacy and magic of matter", but then Pasquero and Poletto can demonstrate with facts and figures that their algae-powered artificial trees are photosynthetic machines many times more efficient than natural ones, while photosynthesis in vitro can be digitally customised with micrometric precision and deliver air filtration to boot. And the reason why slime moulds can serve as oddly reliable problem-solving machines does not reside in any biological black box. Rather, their infra-logical mode of thinking is noticeably similar to various problem-solving strategies that, since 1956, have been at the core of artificial intelligence. Logical thinking is a human prerogative, rooted in the physiology of our neural system. Both slime moulds and machine-learning AI cope without it, so it should be unsurprising that the heuristic premises of contemporary machine learning and gradient-based optimisation are similar in spirit to the 'foraging' problem-solving methods that serve slime moulds so well. I still think that, as far as the mere speed of delivery goes, electronic computing may have a marginal edge over slime moulds; but these are subtleties. In purely epistemological terms, the parallel between monocellular moulds and AI is eloquent and meaningful.

Be that as it may, there is one thing on which we all agree, and which the work of ecoLogicStudio so vividly embodies and represents: in the world of yesterday—from Aristotle to parametricism, with few exceptions—technology was mostly intended to mechanise nature; whereas today's technology has at long last begun to naturalise machines, and with that, hopefully, the way we live, work, and build. And as yesterday's science is evidently leading to, and ending in, disaster, the quest for a 'new kind of science' is no longer a more or less opinable option for our future. It may well be our gateway to the only future we have.

BIOGRAPHIES

Claudia Pasquero is an architect, curator, author and educator. Her work and research operate at the intersection of biology, computation and design. She is co-founder of ecoLogicStudio in London, Professor of Landscape Architecture and Departmental Head of Urban Design at Innsbruck University and Associate Professor at the Bartlett UCL in London. Claudia holds a PhD in architecture and urban design from the Estonian Academy of Arts. She has been Head Curator of BioTallinn the Tallinn Architecture Biennale 2017, and was featured in the WIRED Smart List in the same year. She co-authored *Systemic Architecture: Operating Manual for the Self-Organizing City* with Marco Poletto, published by Routledge in 2012. Claudia is a mother of two and lives in London and Innsbruck.

Marco Poletto is an architect, author and innovator based in London. He is co-founder and Director of the design innovation practice ecoLogicStudio and the PhotoSynthetica venture, focused on developing architectural solutions to fight climate change. His work has been exhibited internationally and is included in the permanent collections of the Centre Pompidou in Paris, the FRAC in Orléans and the ZKM in Karlsruhe. Marco holds a PhD from RMIT University, Melbourne, and co-authored *Systemic Architecture: Operating Manual for the Self-Organizing City* with Claudia Pasquero, published by Routledge in 2012. He currently lectures at the University of Innsbruck and the IAAC in Barcelona. Marco is passionate about long distance cycling and off-road bikepacking.

Sir Peter Cook (b.1936) is a graduate of the Bournemouth School of Art and the Architectural Association, London. He is an architect, lecturer, writer, and curator and was a founder member of the English experimental group Archigram where his projects included the celebrated Plug-in City and Instant City. With various associates he has built the Graz Kunsthaus, housing in Berlin and Madrid and university buildings in Vienna, Gold Coast and Bournemouth, among others. He has been Chair of The Bartlett School of Architecture, UCL and director of London's ICA, and has curated major international exhibitions including the British Pavilion at the Venice Architecture Biennale and Centre Pompidou, Paris. As a writer he has published numerous influential books on architecture since the 1960s. He is a Royal Academician and a Commandeur de l'Ordre des Arts et des Lettres of the French Republic, and recipient of the Royal Institute of British Architects Gold Medal award. He is Emeritus Professor of the Royal Academy, UCL and the Frankfurt Stedelschule. In 2007 he was awarded a knighthood by Queen Elizabeth II for his services to architecture. He continues to practise with CHAP, of which he is a director.

Mario Carpo is an architectural historian and critic. He is Reyner Banham Professor of Architectural History and Theory at the Bartlett, University College London and Professor of Architectural Theory at Die Angewandte (University of Applied Arts), Vienna. He is the author of several books including *Architecture in the Age of Printing* (2001) and *The Second Digital Turn: Design Beyond Intelligence* (2017).

Emmanouil Zaroukas is an architect, researcher and educator. His research interest is focused on the creative capacities of artificial neural networks and machine learning in design. Emmanouil is a lecturer at the Bartlett School of Architecture where he coordinates the History and Theory Module of BPro MArch Urban Design and also teaches histories and theories related to computational processes in multi-scalar built environments. Emmanouil runs his own architectural practice in Greece.

Maria Kuptsova is an architect, artist and researcher. She explores synthetic forms of intelligence and aesthetics, creating cyborganic objects, systems, processes and interfaces. Maria is a graduate from IAAC Institute of Advanced Architecture of Catalonia in Barcelona, where she has continued to work and teach. Maria is a PhD candidate and researcher at the Synthetic Landscape Lab at the University of Innsbruck, as well as collaborating with ecoLogicStudio in London.

Terezia Greskova is a Slovak designer, artist and a yogi. Her research stems from recognising human thought as a relevant instrument for operating within an existing environment. She is currently a PhD candidate and Senior Lecturer at the Synthetic Landscape Lab at the University of Innsbruck. She graduated from the Academy of Fine Arts in Bratislava (Bc) and Vienna (MArch). Terezia has been collaborating closely with ecoLogicStudio since 2015.

Oscar Villarreal is a Mexican architect focused on the exploration of post-anthropocentric architecture and urban landscapes. He holds a Master's degree in Architecture and Urban Design from The Bartlett School of Architecture, UCL where he graduated with distinction and was awarded a gold medal for his dissertation project. Oscar has worked at several architecture studios and founded his own practice in Mexico in 2017.

CREDITS

Projects 1.01. Photosynthetica test beds, pilot projects and early adoptions.

1.01.01 BioBombola
Project by: ecoLogicStudio (Claudia Pasquero and Marco Poletto)
Project Team: Claudia Pasquero, Marco Poletto with Georgios Drakontaeidis, Riccardo Mangili, Eirini Tsomokou.
Academic Partner: Synthetic Landscape Lab, Innsbruck University.
Special thanks to Giacomo and Lulu.
Photography: NAARO.

1.01.02 BIT.BIO.BOT
Project by: ecoLogicStudio (Claudia Pasquero and Marco Poletto)
Project Team: Claudia Pasquero, Marco Poletto with Eirini Tsomokou, Oscar Villarreal, Claudia Handler, Korbinian Enzinger, Terezia Greskova, Alessandra Poletto, Emiliano Rando.
Academic Partners: Synthetic Landscape Lab, Innsbruck University, Urban Morphogenesis Lab, The Bartlett UCL.
Glass 3D printing: Swarovski.
With the generous support of: Innsbruck University, Swarovski, Ecoduna, Destination Wattens, anonymous donor.
Special thanks to the curatorial team at the Venice Biennale: Hashim Sarkis and Gabriel Kozlowski.
Photography: Marco Cappelletti.

1.01.03 Storytelling Bio.Curtain
article by Terezia Greskova.
PhD Candidate. Supervision: Prof. Claudia Pasquero. Synthetic Landscape Lab, Innsbruck University; Dr. Marco Poletto, ecoLogicStudio.
Special thanks to all the team at University of Innsbruck, IOUD, Synthetic Landscape Lab.

1.01.04 Bio.Tech Hut
Project by: ecoLogicStudio (Claudia Pasquero, Marco Poletto).
Project Team: Claudia Pasquero, Marco Poletto with Terezia Greskova, Malte Harrig, Konstantinos Alexopoulos, Apostolos Marios Mouzakopoulos.
Artworks: H.O.R.T.U.S. by ecoLogicStudio (Claudia Pasquero, Marco Poletto).
Structural Engineering: Format Engineers.
General concept: KCA International Designers.
General contractor: Kunzberg, ADUNIC.
Commissioner: Astana EXPO 2017.
Special thanks to Zaure Aitayeva.
Photography: NAARO.

1.01.05 PhotoSynthetica Dublin
Project by: ecoLogicStudio (Claudia Pasquero and Marco Poletto).
Projects team: Claudia Pasquero, Marco Poletto with Konstantinos Alexopoulos, Nico Aulitzky, Shlok Soni, Robert Staples, Chrysi Vrantsi, Chia Wei Yang.

Client: Climat Kic.
Structural Engineering: Nous Engineering, UK.
Bioplastic Supply and Manufacturing Support: Polythene, UK.
Microalgae Cultures Supply: Dr. Fiona Moejes (Bantry Marine Research Station, Ireland).
Photography: NAARO.

1.01.06 PhotoSynthetica Helsinki
Project by: ecoLogicStudio (Claudia Pasquero and Marco Poletto).
Project team: Claudia Pasquero, Marco Poletto with Konstantinos Alexopoulos, Terezia Greskova, Emiliano Rando, Riccardo Mangigli.
Academic Partner: Synthetic Landscape Lab, Innsbruck University.
Client: Helsinki Fashion Week.
Structural Engineering: Nous Engineering, UK.
Biological Medium: Ecoduna.
Photography: Tuomas Uusheimo.

1.01.07 BioFactory
Project by: ecoLogicStudio (Claudia Pasquero, Marco Poletto).
Design Team: Claudia Pasquero, Marco Poletto with Korbinian Enzinger, Claudia Handler, Alessandra Poletto, Emiliano Rando, Eirini Tsomokou.
Academic Partners: Synthetic Landscape Lab IOUD Innsbruck University, Urban Morphogenesis Lab, the Bartlett UCL.
Client: Nestlé Portugal.
Structural Engineer: YIP London.
Biological Medium: Algomed.
Steel Structure: GV Filtri.
Sensory System: ecoLogicStudio.
Visuals: ecoLogicStudio.
Photography: André Cepeda.

1.01.08 AirBubble Playground
Project by: ecoLogicStudio (Claudia Pasquero, Marco Poletto).
Project Team: Claudia Pasquero, Marco Poletto with Eirini Tsomokou, Korbinian Enzinger, Riccardo Mangili, Georgios Drakontaeidis, Terezia Greskova, Stephan Sigl, Alessandra Poletto.
Academic Partners: Synthetic Landscape Lab IOUD Innsbruck University, Urban Morphogenesis Lab, the Bartlett UCL.
Client: Otrivin, GSK.
Structural Engineer: YIP London.
Biological Medium: Ecoduna.
Timber Structure: ArchWood.
Membrane: Temme Obermeier.
Special thanks to Farhad Nadeem, global brand director at Otrivin, Haeleon and Lauren Dyer, creative director at Saatchi & Saatchi.
Photography: Maja Wirkus.

Projects 1.02. Bio-digital sculptures: Designing the living.

1.02.01 SuperTree, ZKM, Karlsruhe
Project by: ecoLogicStudio (Claudia Pasquero, Marco Poletto).
Project team: Claudia Pasquero, Marco Poletto with Andrea Bugli, Terezia Greskova, Chiara Catalano, Kiki Goti, Nikos Xenos, Alberto Chiusoli.
Academic Partners: The Architectural Association London, The Bartlett UCL, The Aarhus School of Architecture.
In memory of the late Peter Weibel and with special thanks to Philipp Zeigler
Photography: NAARO, ecoLogicStudio.

1.02.02 H.O.R.T.U.S. XL Astaxanthin.g
Commissioner: Centre Pompidou, Paris.
Exhibited to date at: Centre Pompidou, Paris. MAK Museum, Vienna. Mori Gallery, Tokyo. Fundacion Telefonica, Madrid. Hyundai Motor Studio, Busan.
Project by: ecoLogicStudio (Claudia Pasquero, Marco Poletto).
Project Team: Claudia Pasquero, Marco Poletto, Konstantinos Alexopoulos, Matteo Baldissarra, Michael Brewster, Maria Kuptsova, Terezia Greskova, Emiliano Rando, Jens Burkart, Niko Jabadari, Simon Posch.
Academic Research Partners: Synthetic Landscape Lab, IOUD, Innsbruck. University; CREATE Group, WASP Hub Denmark, University of Southern Denmark SDU; The Urban Morphogenesis Lab, The Bartlett UCL.
Engineering: YIP structural engineering.
Microalgal Medium Material Support: Ecoduna AG, Euglena Co.
3D printing Material Support: Extrudr.
Special thanks to curators Marie-Ange Brayer, Frédéric Migayorou. Olvier Zeitoun and to Innsbruck University for the support.
Photography: NAARO, Kioku Keizo, Peter Kainz.

1.02.03 Arbor
article by Maria Kuptsova.
PhD candidate, Supervision by: Prof. Claudia Pasquero, The Synthetic Landscape Lab at Innsbruck University, Austria.
Special thanks to all the team at Innsbruck University, IOUD, Synthetic Landscape Lab.

1.02.04 bl.O.serie
Project by: ecoLogicStudio (Claudia Pasquero and Marco Poletto).
Project Team: Claudia Pasquero, Marco Poletto with Eirini Tsomoku, Joy Boulois, Claudia Handler, Emiliano Rando.
Academic Partners: Synthetic Landscape Lab at Innsbruck University, Urban Morphogenesis Lab at the Bartlett UCl.
Commissioner: German Cultural Foundation, Kommunikation Museum Frankfurt, Berlin.
Special thanks for Innsbruck University for the support.
Photography: Sven Moschitz, MSPT.

1.02.05 meta-Folly
Project by: ecoLogicStudio (Claudia Pasquero and Marco Poletto).
Project Team: Claudia Pasquero, Marco Poletto with Andrea Bugli, Philippos Philippidis, Mirco Bianchini, Fabrizio Ceci, Phil Cho, George Dimirakolous, Manuele Gaioni, Giorgio Badalacchi, Antonio Mularoni, Sara Fernandez,

Daniele Borraccino, Paul Serizay, Maria Rojas, Anthi Valavani.
Commissioner: FRAC collection _Orléans _France.
The pavilion is now part of the permanent collection of the FRAC Center.
Special thanks to curators Frédéric Migayorou, Marie-Ange Brayer.
Photography: ecoLogicStudio.

Projects 1.03. Cyber-gardening the city.

1.03.01 Urban Algae Canopy
Project by: ecoLogicStudio (Marco Poletto, Claudia Pasquero).
Project Team: Claudia Pasquero, Marco Poletto with Andrea Bugli, Elisa Bolognini, Alessandro Buffi, Julien Sebban.
Academic partners: The Bartlett UCL, Aarhus School of Architecture.
Digital responsive systems: Alt N – Nick Puckett.
Structural Engineering: Ing. Mario Segreto, Ing. Nicola Morda.
Project Management: Arch. Paolo Scoglio.
ETFE contractor: Taiyo Europe GmbH.
Timber contractor: Palumbo Legnami.
Microalgae culture: Algain Energy srl.
Special thanks to EXPO Milano 2015 Future Food District curator Prof. Carlo Ratti, director at the MIT Senseable City Lab, to the area manager Matteo Gatto.
Photography: ecoLogicStudio.

1.03.02 Urban Algae Folly
Project by ecoLogicStudio (Claudia Pasquero and Marco Poletto).
Project Team: Claudia Pasquero, Marco Poletto with Konstantinos Alexopoulos, Apostolos Marios Mouzakopoulos, Matteo Pendenza.
Academic Partners: The Bartlett UCL, the Aarhus School of Architecture.
Structural Engineering: Ing. Mario Segreto, Ing. Nicola Morda.
Project Management: Arch. Paolo Scoglio.
ETFE contractor: Taiyo Europe GmbH.
Metal Structure: GV Filtri.
Microalgae supplies: Algain Energy srl.
Clients: City of Braga, City of Aarhus.
Special thanks to Claus Peder Pedersen for his precious support and Marco's PhD supervision. Also our gratitude goes to RMIT and Marco's Melbourne based PhD supervising team: Marcelo Stamm, Paul Minifie and Roland Snooks.
Photography: NAARO, Claus Peder Pederson, ecoLogicStudio.

1.03.03 AirBubble COP26
Project Name: AirBubble air-purifying eco-machine.
Project by ecoLogicStudio (Claudia Pasquero, Marco Poletto).
Project Team: Claudia Pasquero, Marco Poletto with Greta Ballschuh, Sheng Cao, Korbinian Enzinger, Claudia Handler, Riccardo Mangili, Alessandra Poletto, Eirini Tsomokou.
Academic Partners: Synthetic Landscape Lab IOUD Innsbruck University, Urban Morphogenesis Lab BPRO The Bartlett UCL.
Client: Otrivin®.
Structural Engineer: YIP London.

Biological Medium: Algomed.
Pneumatic Structure: Pneumocell.
Sensory System: ecoLogicStudio, Puckett Research, Almondo.
Special thanks to Farhad Nadeem, global brand director at Otrivin, Haleon.
Photographer: ©NAARO.

Projects 2.01. Synthetic landscapes.

2.01.01 Les Jardins Fluviaux de la Loire
Project by ecoLogicStudio (Claudia Pasquero and Marco Poletto).
Project Team: Marco Poletto, Claudia Pasquero with Konstantinos Alexopoulos, Matteo Pendenza, Mauro Mosca, Apostolos Marios Mouzakopoulos, Gabriela Zarwanitzer.
Academic Partner: Urban Morphogenesis Lab at the Bartlett UCL.
Special thanks to Rc16 students and teaching faculty for their contribution.
Commissioned: for Biennale d'Architecture d'Orléans#1, as part of the permanent collection of FRAC Center, by curators Abdelkader Damani and Luca Galofaro.
Photography: ecoLogicStudio.

2.01.02 Solana Open Aviary
Project by ecoLogicStudio (Claudia Pasquero and Marco Poletto).
Design Team: Marco Poletto, Claudia Pasquero, Terezia Greskova, Vlad Daraban.
Academic Research Partner: Urban Morphogenesis Lab at the Bartlett UCL. (Claudia Pasquero, Maj Plemenitas, Stuart Maggs), Arkitectskolen Aarhus (Marco Poletto), European Space Agency.
Commissioned: for the Montenegro Pavilion at the Venice Biennale 2016.
Special thanks to curators Bart Lootsma and Katharina Weinberger, commissioner Dijana Vucinic on behalf of the Montenegro Ministry of Sustainable Development and Tourism.
Photography: ecoLogicStudio.

2.01.03 The Anthropocene Island project
Project by ecoLogicStudio (Claudia Pasquero and Marco Poletto).
Project Team: Claudia Pasquero, Marco Poletto with Konstantinos Alexopoulos, Raphael Fogel, Terezia Greskova, Mauro Mosca, Anjie Gu, Matteo Pendenza.
Head Curator: Claudia Pasquero.
Artists and Scientists: Alisa Andrasek at RMIT, Maj Plemenitas at Linkscale, Edouard Cabay at Appareil, Heather Barnett at Central St Martins, Claudia Pasquero and Marco Poletto at ecoLogicStudio, Noumena, Studio Unseen, Claudia Pasquero at Urban Morphogenesis Lab and Marcos Cruz at BiotA at The Bartlett UCL, Rachel Armstrong and Experimental Architecture Newcastle University, Areti Markopoulo and the IAAC, Veronika Volk and Tommas Tammis at EKA.
TAB 2017 Catalogue and Webzine Editor: Lucy Bullivant.
TAB 2017 Catalogue and Webzine Graphic Design: AKU collective.
Organised by: Estonian Centre of Architecture (MTÜ Eesti Arhitektuurikeskus)
Special thanks to TAB2017 commission and production team, as well as the continuous support of the Estonian Academy of Art in Tallinn.

Our gratitude goes to Claudia's PhD supervisors Veronica Valk.
Photography: NAARO, Tõnu Tunnel.

2.01.04 Biocoenosis Nest: A city woven by collective intelligence
article by: Oscar Villarreal.
MArch Thesis supervised by: Prof. Claudia Pasquero, Filippo Nassetti, Emmanouil Zaroukas, Urban Morphogenesis Lab, UCL, London.
Special thanks to all the team at Bartlett BPRO, University College London.

2.01.05 GAN.OS 101
article by Emmanouil Zaroukas.
History and Theory tutor.
MArch Thesis by Anshika Tajpuriya, Shusheng Huang, Meng Zheng, Tao Chen, Sheng Cao.
Supervised by Prof. Claudia Pasquero, Filippo Nassetti, Emmanouil Zaroukas at RC16, the Urban Morphogenesis Lab at the Bartlett UCL, London.
Special thanks to Emmanouil Zaroukas (Manos) for the intellectual contribution to not only this article but most of the text presented in this book, and to all the team at Bartlett BPRO, University College London for the support.

Projects 2.02. Deep planning.

2.02.01 GAN-Physarum: la dérive numérique
Project by: ecoLogicStudio (Claudia Pasquero and Marco Poletto)
Project Team GAN-Physarum: la dérive numérique. Biopainting 2022:
Claudia Pasquero, Marco Poletto with Greta Ballschuh, Sheng Cao, Alessandra Poletto.
Project Team GAN-Physarum: la dérive numérique. AI video:
Claudia Pasquero, Marco Poletto with Joy Boulois, Korbinian Enzinger, Oscar Villarreal.
Project Team Deep Green Urbansphere 2021:
Claudia Pasquero, Marco Poletto with Thole Althoff, Michael Brewster, Xiaomeng Kong, Stephan Sigl, Eirini Tsomoukou, Lixi Zhu.
Academic Partners: Synthetic Landscape Lab at Innsbruck University, Urban Morphogenesis Lab at the Bartlett UCL.
Commissioned by: Marie-Ange Brayer and Olivier Zeitoun, Centre Pompidou.
Special thanks to Frédéric Migayorou for the continuous support of the research on *Physarum polycephalum*.
Permanent collection: Centre Pompidou Paris.
Photography: NAARO.

2.02.02 DeepGreen, the case of Guatemala City
Project by: ecoLogicStudio (Claudia Pasquero and Marco Poletto).
Design team: Thole Althoff, Michael Brewster, Xiaomeng Kong, Stephan Sigl, Eirini Tsomoukou, Lixi Zhu.
Academic Partners: Synthetic Landscape Lab at Innsbruck University, Urban Morphogenesis Lab at the Bartlett UCL.
Commissioned by: United Nations Development Programme (UNDP).
Special thanks to Lejla Sadiku, UNDP project leader.

INDEX

Note: Illustrations are indexed in *italic* page numbering.

A

Aarhus, Denmark 188, *203*, 205, 208, *210*, 212–13, 215
Aarhus Wet City Project 204–5, 214–15
actuating systems 30
Adamatzky, Andrew 232, 240, 244
aeration systems 27–8, 31–2, 34, 107, 225
The Aesthetic Brain 237
aesthetics 232–4, 340; of nature 25, 232–4; polycephalum 233; unrivalled 26
agri-territories 185; *see also* farmlands
agriculture 182, 195, 199, 254, 326, 333–4
air filtration systems 100, 338
air molecules 28
air piping systems 42
air pollution 14, 19, 28, 40, 44, 79, 98, 102–3, 146, 159, 194–5; filtering performance 34; global urban 13; harmful 20; monitoring 33; particles 143; urban 35, 103
air purification 12, 18, 26, 43
air quality 13–14, 19, 32–5
Air Quality Index 14, 32–3, *99*
air recirculation 99, 104, *222*
AirBubble Project 19, *97*, 98–9, *100*, 102–3, *104*, *105*, *107*, *108*, *217*, 218, *220*, *221*, *223*, *224*, *226*; air purifying eco-machine 218–19, 222; eco-machine installed at the Glasgow Science Center *220*, 225; exhibition at the Copernicus Science Centre, Warsaw *61*; hosts 52 large bioreactors in borosilicate glass 102; invents a new architectural typology 99
algae 5–6, 12–13, 15, 17, 19–20, 22–5, 28, 51, 58, 75, 79, 102, 107; cells 22, 28, 122, 124; cultures 27, 67, 102; flow system 197, 199–200, 211; fresh 24; garden systems 51, 99

algorithms 30, 137, 154, 208–9, 212, 240, 242, 248, 250, 312, 319; agent-based edge tracing 275; architectural 239; computational 189; customized 33, 239; deploying design 239; digital 120; evolutionary neural network 154; generative 233; internal 247; machine learning 150, 319, 326; minimal path 249; novel non-standard 240; scripted 337; trained 319
analogue computers 238
analogue models 249
Anthropocene 112, 166, 236
Anthropocene Island Project 282–3, *284*, *287*, *290*, *291*
ants 187–8
AQI *see* Air Quality Index
Arbor Project *149*, 150, 156
Archilab 9 162, *164*, 170
architects 112, 118, 170, 202, 239, 339–40; Cedric Price 118; Diana Vucinic 266; Frei Otto 243; Johan Fraser 239; pioneering 239
architectural 12–13, 25–6, 38, 50, 57–8, 62, 64–5, 99, 113–14, 117, 119, 121–2, 143, 184, 197–9, 238–9, 272, 295, 299, 339–40; design methods 232; integration 13, 15, 20, 25, 31; partitioning systems 57, 62; retrofitting 38, 76, 84, 88
architecture 2–3, 64, 84, 112–14, 117–24, 161–2, 166–7, 186–8, 198–9, 201–2, 213, 232–3, 238–9, 249–50, 267, 339–40; advanced 46, 340; algorithmic 306; animate 238; bio-computing 237; bio-digital 34, 143, 317; biotechnological 122–3; of carbon neutrality 13, 217; computational 238, 240; ecological 232; innovative 197–8; photosynthetic 58; polycephalum 232–3, 241; post-anthropocentric 340; simulated 187; traditional 315
areas 115, 146, 214, 246, 262, 332; agricultural 181, 185; green 190; metropolitan 181, 183; park 205; rural 258; urban 185

347

artificial intelligence 2–3, 92, 112–13, 122, 124, 235, 242, 250, 306, 322, 324; of autonomous farming protocols 89; biologically augmented 30; contaminated habitats and ubiquitous forms of 112; core of 338; and the metaverse 235; non-human 315; and synthetic biology 235; ubiquitous forms of 112
artificial landscapes 267–8, 275; *see also* landscapes
Aztecs 16, 24

B

Back to Future Exhibition *160*
Barrett, Lisa Feldman 62
Bartlett School of Architecture 232, 267, 339–40
Bateson, Gregory 234, 234n2, 234, 237; anthropologist and cybernetician 232; argues we live in a world populated by ecologies of mind 233; and the relationship between the human and the non-human realms 236; writes *Steps to an Ecology of Mind* 233
Bateson, M.C. 119n7
Batty, Mike 186–7, 189
behaviour 58, 114, 121, 188, 237, 240–1, 245–7, 249, 262, 270; adaptive 14; collective 59, 115, 137, 187, 245; complex migrating 276; emergent complex 170; non-linear 255; oscillatory 241; patterns of 154, 318–19; reinterpreting Physarum Polycephalum's 319; slime mould 295
Bergson, Henry 237–8
Bio.Bombola Project 2, 6, 28, 51
bio-computing architecture 237
bio-design 2, 92, 298, 302
bio-digital age 276
bio-digital architecture 34, 143, 317
bio-digital cultivation 120, 195
bio-intelligence 98
bio-painting 249, *309*, *310*, 311
bio-pixels 137, 142
bio-printing 154
bio-reactors 120
Bio.Tech Hut Project 6, 28, 64–5, *66*, 68, 70, *73*, 120, 146
BioBombola Project *39*, 40, *42*, *43*, 44
Biocoenosis Nest *293*, 294–5, *296*
BioFactory *88*, 89, 92
biofuel 24, 69–70, 195, 301
biological intelligence 47, 89, 121, 142–3, 150, 155, 318, 321
biological models 136, 241, 295
biological structures 150, 244
biomass 12, 16, 19, 23, 47, 51, 65, 68–9, 133, 142, 146, 190, 193; algal 14, 23; extracted 23; fibrous 84; harvested 23, 88; microalgae 17; new 146; production 14, 18, 195
bioplastics 14, 23–7, 76, 79
biosphere 59, 79, 84, 112, 180–2, 184–6, 189, 202, 235–6, 243–4, 249
BioTallinn 282, 339
biotechnological 19, 32, 110, 242; architecture 122–3; playgrounds 12, 35, 97; re-metabolisation 289; stratum 194; system 12
biotic molecules 263
biotic network 332–3, 336
BIT.BIO.BOT Project 2, 6, 47, 50
bloodstream 32–3
Boeri, Stefano 194
Bogatyreva, Vincent 150
botanists 185
brain 6, 33, 240–2
buildings 13–14, 18, 68, 84, 119, 143, 146, 167, 190, 215, 222; bio-factories 87–8; and cities 68; designing of 76; facades 12, 92; new 14; university 339; warehouse 25
Burdett, Richard 194
Busquets, Joan 194
buzzers 163, 173, 175; piezo-triggered analogue 163, 170; piezoelectric 174; set to mimic a swarm of crickets in a field 170

C

Cache, Bernard 241
canopy *131*, 132, *196*, 198–200
carbohydrate molecules 17
carbon 12–13, 18–19, 22, 24, 26, 63, 68, 77, 84, 88, 133; dioxide 5, 12, 15, 17–18, 40, 59, 65, 75, 84, 102, 159; footprint 235; neutrality 5, 12–13, 35, 51, 68, 70, 75, 80, 183,

217; sequestration 13, 17, 20, 28, 70, 79, 195
Carpo, Prof. Mario 4, 337, 340
cells 13, 16, 19, 21, 28, 112, 114, 137, 158, 160, 247; algae 22, 28, 122, 124; engulfing 115; eukaryotic 114; Eukaryotic 114; floating 22; individual 121, 137, 142, 159; inflatable 219; inhabitable bioreactor 283, 289; living Chlorella SP 158; micro-algae 54, 159; motile 16; photosynthetic 22; printed reactor's 159; prokaryotic 115; Spirulina 21, 54
cellular aggregation of a virtual colony of cells 158, *160*
Centre Pompidou, Paris 120, 136, *140*, 306–7, *308*, *309*, 317, *318*, 339
change 4, 16, 19, 34–5, 84, 181–3, 188, 190, 233, 235, 237–9, 246, 248, 314, 326–7; behavioural 35; creative 319; positive 327; rapid 84; significant 190; technological 306; *see also* climate change
Chatterjee, Anjan 237
chemical energy 17
children 2, 5, 13, 40, 98, 102–3, 218
Chlorella cultures 52, *100*, 218–19, *223*
Chlorella SP 47, *48*
chlorophyll 21
cities 12–13, 68, 102–3, 121–2, 178–92, 204–5, 208–9, 236, 238–9, 248–9, 253–5, 258, 262–3, 294–5, 315, 322–3, 333–4; bio-conscious 12; carbon neutrality in 5; contemporary 184, 187; cyber-organic 155; decarbonising 12; European 185; fictitious 311; fractal 240; fractal nature of 189; global 13, 183, 311; gridded 311; liquid 186; marshland 295; modern 258; neutral 12, 18, 190; polluted 16, 98; re-greening 236; small 254
civilization 181–2, 235
Clarks, Andy 232, 242
Clement, Gilles 118
climate change 4, 19, *76*, 233, 235, 295, 323, 326, 339; catastrophic 4, 233; fighting 339; global 76, 235, 323; scenarios 295
Climate Innovation Summit 2018 76
closed loops (algae flow) 199–200

co-evolutionary 202, 237, 244, 262, 265, 270, 273, 330; blue-green urban plans 212; designing systemic models 188; systems 5, 234, 237; with urban infrastructures 205
CO2 emissions 12–13, 15, 18–19, 22–3, 27, 68, 70, 76, 120, 129, 133, 204
cognitive systems 232, 248, 311
collective behaviour 59, 115, 137, 187, 245
collective intelligence 3, 57–8, 62, 187, 252, 294–5
colonies 18, 115–16, 158, 188; ant 187–8; coral 114–17, 120, 137; of cyanobacteria 132, 137; living algae 68, 299; living microalgae 12; photosynthetic 65; urban bacteria 133
coloured circles 34–5
colours 16, 21, 29, 34, *171*, 198, 200, *287*, 319, 327; and the darkening of the canopy 199; and the distribution on the substratum 248; indexing different stages of the wastewater treatment and pathogen density 287; and the nuances in real time 247; of the slime mould 3, 233, 240–2, 244–50, 296, 311, 314–15, 338
communication 155, 187, 205, 244, 250; cybernetic 234; devices 233; direct 277; local 188; meta-logical 233; non-verbal 232; renewed 237; stigmergic 187; systems 188
computation 118, 195, 238–41, 245, 250, 339; analogue 234; design 47, 120, 123, 166, 238–9, 241, 295; designers 241; evolving collection of spatial protocols of 112; grid 190, 262–3; processes 236, 247–8, 340
computers 238–40, 244; analogue 238; biological 245, 312; general-purpose 244
computing 155, 188, 241–2, 249–50, 295
contemporary architecture *see* architecture
contemporary design *see* design
contemporary machine learning 338; *see also* machine learning
Convivium *50*, *51*, *53*, 54, *55*
Cook, Sir Peter 4, *5*, 118, 339
COP26 13, *192*, 217–18, 220

Copernicus Science Centre, Warsaw 61, *104*, *105*
coral colonies 114–17, 120, 136–7
Corderie dell'Arsenale 46, *48*, *49*, 50, *53*
cosmetics 14, 16–17, 24
crickets 162, 170
cultivation 15–16, 25, 54, 64, 113, 121, 195, 198; algae 40; bio-digi 120, 195; collective urban 210, 213; domestic 46; high-tech 44; indoor 40; practices of 112; processes 195, 198; small-scale 202
cultures 20–3, 26–31, 33, 65, 199, 203–4, 210, 242, 258; algae 27, 67, 102, 203; biological 30; Chlorella 52, 100, 218–19, 223; cyanobacteria 128, 130; living 15, 20, 25, 31–2, 34, 60, 68, 70, 159, 198, 204; micro-algae 15, 50–1; microalgal 76, 146; microbial 120; photosynthetic cyanobacteria 120, 142; Spirulina 22, 34, 197
cyanobacteria 15, 17, 42, 63, 65, 68, 120, 128–30, 198, 233; cultures 128, 130; prokaryotic 17, 120; species of 16
cyber-gardens 4, 12, 120, 192
cyberneticians 118, 232; Gordon Pask 118, 232; Gregory Bateson 232, 234, 237
cybernetics 119, 149–50, 166, 234, 239

D

Da Vinci, Leonardo 183
DeepGreen urban design paradigm 117, *304*, *321*, 322, 326–7, 334
Deleuze, G. 114, 114n1, 311
design 2, 4, 26, 31, 112, 114, 150–1, 155, 186–7, 232–3, 236, 249, 339–40; bio-digital 62, 213; biospheric 242; computational 47, 120, 123, 166, 238–9, 241, 295; contemporary 114; decisions 31, 190; integrated 273; intelligence 2–4; optimal 33; practice 3, 232, 319; problems 239, 249; processes 117, 189, 233, 239, 249, 262, 334; projects 188, 232, 276; of resilient cities 326; solutions 4; strategic 20; trans-disciplinary 276; urbanspheric 186

design innovation 2–3, 15, 98, 113, 121; consortium 2; digital 213; extended framework 232; systems 117
diagrams 31, 33, 35, 77–8, 108, 176, 186; BioBombola 43; emerging 327; PM2.5 pollution monitoring 108; relational 117; single consistent 35; urban 295; and visual maps 31
digital elevation models *324*, 327
digital models 26, 142
direct investments 327
drawings 160, 243, 248–50, 263, 285; analogue 249; emerging 248; exquisite 118; operational field 263; original 123; shop 26
Dublin 76, *77*, 99, 122, 222
Dussutour, Audrey 240

E

eco-machines 218–19, 238
ecological 3, 12, 58, 114, 121–3, 181–4, 212, 218, 232, 234–6, 243; Armageddon 184; catastrophe 182, 235; deficit 243; footprint 212; impact 182; innovation 123; intelligence 3, 12
ecoLogicStudio 2, 24, 34–5, 54, 64–5, 103, 122–3, 188–9, 194, 218–19, 232, 266–7, 277, 282, 338–40; proposal challenges conservative sentiments with a masterplan intended to promote a new urban morphogenesis 283; test beds in Dublin and Helsinki 122; work goes hand in hand with the redefinition of a common term 'prototype' 187
ecologies 59, 118, 121, 232–4, 300; artificial 270; bacterial 270; dark 234; disrupted 234; global 277; living 13; local 270; micro 120, 204; micro-algal 128; non-human 234; novel 236; urban 121, 238; wild plant 274
ecologists 235, 238
ecosystems 112, 204, 234–5, 273, 294, 326; augmented 283; growing 30; living 180; marine 15; planetary 294; re-green global 234; sustainable business 23
emissions 12, 23, 68, 88, 196, 219, 289; CO_2 12–13, 15, 18–19, 22–3,

27, 68, 70, 76, 120, 129, 133, 204; dramatic reduction in 13; and energy consumption 18; monitoring urban 12; zero harmful 68
endosymbiosis 114, 136; *see also* coral colonies
energy 14, 18, 21–2, 64–6, 68–70, 84, 120, 122, 181, 183, 245–8, 272–3, 314; chemical 17; collective 326; consumption 18, 70, 235; efficiency 14; expenditure 188, 244; farms 184; infrastructures 209; light 17, 31; solar 12, 14, 18, 102–3
English landscape gardens (18th century) 167
environment 20, 22, 24, 26, 31, 113–14, 116, 119, 187–8, 215, 218, 242, 245; active 299; artificial 64, 238; changing 30, 116, 123; commercial 68; constructed 315; domestic 42; immersive 167; natural 64, 167, 180; sterile 314; surrounding 5, 34, 42, 59, 84, 113, 151, 188, 226, 237; urban 13–14, 19, 30–1, 34, 132, 180, 185, 187, 198, 212, 302
environmentalism 178, 180–1, 183, 283
ESA see European Space Agency
Estonian Biennale of Architecture (BioTallinn) 282
ethylene tetrafluoroethylene 25, 89, 99–100, 198, 200, 207
eukaryotic cells 114
eukaryotic microalgae 17
European Space Agency 267, 275–7, 288
evolution 2, 84, 99, 114, 180, 197–8, 222, 233, 280, 294, 322; biological 112, 128; cultural 3; experimental 28; social 182; structural 264; technological 235
exposures 32, 115, 132–3; to elevated concentrations of NO2 32; optimal 22; prolonged 14; protracted 32; solar 116

F

fabrication 25–6, 54, 129, 170; files 26; material 6; methods 154; process 26, 219; techniques 47
false colour images (called 'infrared,' 'normalised water' and 'vegetation'

indexes) 277
families 2–3, 17, 39–40, 42, 51
farms 12, 20, 23, 29–30, 48, 51, 89, 92, 181–5, 194, 199
flocculation process 29
flue gasses 19; *see also* gasses
food 14, 16, 19, 24, 26, 54, 56, 65, 68–9, 195, 246–7; distribution 12, 150; industry 24, 95; production systems 24, 301; sources 16, 314; supplements 16; unhealthy 13
FRAC Centre, Orléans 123, 162, 164, 170, 240, 253–63, 339
Fraser, Johan 239
'Fun Palace' Project 118
Future Food District *196*, 198
Futurium, Berlin *130*

G

Galvanic Skin Response 59
GAN *see* Generative Adversarial Networks
GAN-Physarum 241–2, 250, 304, 306, 308–9, 311, 315, 319, 327; algorithm 307, 313, 316–17, 320, 329; and machine vision 34; projects an input image in the biological domain A to an image in the urban domain B 319; reimagines Guatemala City 334; self-development of 319; value of 311; workflow 319
GAN.OS *see* Geological Adversarial Network of Settlements
Garden Hut 65, *69*, 74
Garden of Accretion *256*, 259, 263–4
Garden of Entanglement. *259*, 263
Garden of Sedimentation *259*
gardens 12, 65, 118, 259, 262–3; biotechnological 194; domestic algae *39*, 40; English landscape 167; new public fluvial 253; personalised virtual 212; physical 262; sedimentary 258; urban 212, 263; vertical 13, 18, *49*, *50*, 51; virtual botanical 212
gasses 12, 19–20, 32
gastrodermis 115
Generative Adversarial Networks 151, 154, 233, 298–9, 301–2, 306, 315, 318
generative models 249, 306

'Generator' Project 118
generators 118, 318
Geological Adversarial Network of Settlements 298, *299*, *301*, 302
GIS data 31, 35, 327
GIS maps *325*
Glasgow Science Centre 218, 220
Global Footprint Network 243
gradients 121, 154, 245–6; articulated 269; atmospheric 289; Habitat Gradients Operational plan 269; of light 246
Graeber, David 181–3
Graphic User Interface 33
green microalgae 16, 70
Green Urban Networks 215
Grešková, Terézia 58
growth protocol 20, 23, 30
growth rate 19, 21–2
GSR *see* Galvanic Skin Response
Guatemala City 304, 322, 324, 326, 330, 334–6
Guatemala City Blue-Green Masterplan 325, *328*, *329*, *331*
Guattari, Felix 114, 121, 311
GUI *see* Graphic User Interface

H

Habitat Gradients Operational Plan 269
Haematococcus pluvialis (fresh water red microalgae species of Chlorophyta) 17
harvesting systems 27–8
health 34, 100, 103, 295; children's 102–3; crises 14, 44, 235; global threats 98; human 20, 32–3; problems 32; risks 35
healthcare 272, 276
Herzog, Jacques 194
heterogeneous systems 123, 155, 238
H.O.R.T.U.S. XL 120–1, 136, *138*, *140*, *141*, 143
human intelligence 242, 250, 294, 298
human perception 4, 58–9

I

individual cells (bio-pixels) 137, 142
informational networks 205, 244, 283
infrastructure 187, 195, 209, 242, 248, 276, 318, 323; bio-smart 98; current biotic 213; efficient wet 189; energetic 242–3; energy 209; green 208; human 244; living organic 329; new 98, 194; original 183; seawater management 272; simulating traffic 245; urban 112, 122, 124, 180, 189, 205, 212, 243, 248, 255, 263; waste 329–30, 334–5
installations 5–6, 31, 46, 54, 59, 84, 89, 113, 117, 119–20, 219
intelligence 3, 112, 117, 119, 122, 124, 233, 235–6, 241–3, 298, 300, 323, 326, 340; alien 241; animal 122, 143; artificial 2–3, 89, 92, 112–13, 122, 124, 235, 242, 306, 322, 324; bio-artificial 136, 239; bio-digital 123; biological 121, 142, 150, 155, 318, 321; collective 3, 57–8, 62, 187, 252, 294–5; computational 30; design 2–3; digital 306; distributed 235, 246; ecological 3, 12; embedded 34, 118; non-human 235; spatial 121
Intergovernmental Panel on Climate Change 13
interventions 183, 233, 255; bio-architectural 285; direct human 30, 263; ecoLogicStudio's 270; on-site 277
Inuit 182
investments 23, 327, 334
IPCC *see* Intergovernmental Panel on Climate Change
Italian Renaissance 183
Italy 181, 184

J

Japan 141, 240
Jardins Fluviaux 252, 254, *259*

K

Kamchatka *299*
Karlsruhe 54, *128*, 339
Kazakhstan *64*, *74*, 120
knowledge 28, 31–2, 59, 112, 123, 237, 255, 280, 300, 302; architectural 64; conceptual 237; cultural landscape synthesising 254; curate spatial 112; human 298; production 302
Kolmogorov-Uspensky machine 244

Kunick, Wolfram 185
Kuptsova, Maria 150–1, 340

L

Ladybug Grasshopper plug-in 31
landscapes 3, 113, 120, 183, 246, 248–9, 252, 254, 267–8, 275–7, 283; artificial 267–8, 275; biomechanical 118; cognitive 248; culinary 53–4; cultural 255, 258; design 254; ecology 273; emergent anthropic 258; existing 288–9; infrastructural 270, 280; local 183, 270; manufactured 248; marshland 296; natural 113, 329; new 117; productive 194; sedimentary 258; synthetic 4, 23, 46, 58, 60, 136, 143, 151, 288, 314, 340; tectonics 275
language 3, 5, 161–2, 170, 186, 232, 234, 242; new architectural 199; schematic 237; typological 189
lasers 31, 132, 169, 175, 207, 278
Latour, Bruno 236
Laugier, Marc-Antoine 64
layers 25, 27, 115, 120, 137, 142, 187, 271, 278, 284, 322–3; active 15; aquatic 284; biotechnological 107; blue-green 205; cognitive 187; fused glass 54; operational 271; thin ornamental 184, 246
Le Duit Saint Charles 263
learning 155, 338, 340
Leonardo Da Vinci 183
Les Jardins Fluviaux de la Loire 254, 259
levels 29, 33, 89, 122, 186, 190, 208, 268, 271, 276–7; concentration 108; high 255; illegal 13; international 327; lighting 31; multiple 65, 114–15; personal 58; reaching non-human 112; salinity 29; scalar 190; sea 299
LGem 65
life 13, 18, 64, 112, 119, 182, 185–6, 233, 237, 262, 311; cycle 123, 158, 314; daily 235, 238; inorganic 238; interpreting 237; new 276; nomadic 311; science 136; span 4
lifestyles 3, 167, 182
light 14, 16, 21, 31, 33, 167, 171, 185–6, 188, 239, 244, 246; artificial 21; diffused 226; energy 17, 31; fields 246; incoming 27, 199; intensity 21; penetration 137; screening 120; sensors 29; visible 65; wavelengths 21
liquidity 186–7
living 4, 12–13, 15, 30–2, 49–51, 64–6, 83–4, 102–4, 117–18, 120–1, 138–40, 154–5, 157–9, 185, 197–8, 218–19, 242–4, 248–50, 294–5, 310–12; algae 68, 299; architectures 117–18, 143; biological organisms 121; Chlorella *138*; Chlorella cultures 138–9, 219, 223; cladding 49–51; cultures 15, 20, 25, 31–2, 34, 60, 68, 70, 159, 198, 204; entities 294–5; microalgae 12; organisms 47, 59, 92, 113, 136, 166, 250; planet 4; systems 118, 166, 185, 233, 237, 283, 288
Living Hut 64–6
locations 31, 52, 102, 162, 189, 212, 246–7, 288, 327; food source 246; geographic 208; mining 247; optimal habitat 295; simulated 259; urban 181
Loire, River 262–3
Loire Valley 252–9, 261–2
loops 27–8, 65–6, 118, 170, 199–200
Lootsma, Prof. Bart 266
Lovecraft, H.P. 298
low-carbon construction process 89

M

machine learning 155, 338, 340; *see also* learning
machine vision 30, 34
machines 119, 169, 233, 250, 270, 277, 311; architectural 318; autonomous farming 122, 143; calculating 239; carbon sinking 128; centrifugal 28; corporate 202; digital 250; human 238; industrial 26–7; metabolic 120; photosynthetic 338; precision 25; printing 120, 142, 246; productive 183; sorting 114; virtual 277
macro-behaviours 122, 124
Maggs, Stuart 267
main facades *77*, *78*, *85*, *86*
maintenance 26; costs 18, 23; high 13; low 17; regular 30; subcontracting 23

Malafouris, L. 59
Malmo 190
management 15, 118, 212, 255, 263
maps 186, 208–9, 310, 315; background 248; barometric weather 189; global 254; infrared 275; intensive urban field 189; operational field 277; territorial 246
masterplans 187, 194, 271, 283; blue-green 205, 282, 285; original 183; regional 35; top-down 188
material systems 154, 187
McDonough, William 194
media 3, 167, 235, 248, 263, 300
membrane 107, 206, 221, 224, 245; empty 219; inner 224; inverted conical roof 99, 104; living 84; outdoor 222; single stretchy 314; technology 25
memory 59, 243, 245, 247; distributed spatial 245, 247–8; embedded 248; outsourced 248; slime mould's spatial 246
Menges, Axel 188, 244n19
Mesoamericans 16
meta-languages 3, 161–2, 170, 186, 234; *see also* language
metabolism 29, 47, 112, 128, 142, 243, 247; accelerated 19; circular 62; complex 18; expanded 204; fluctuating 112; human 128; increasing 183; internal 289; slime mould's 246; urban 195, 281–2
Metropolitan proto-Garden 194
Mexico City 24, 334, 340
micro-organisms 66, 113, 236
microalgae 12, 14–20, 24, 28, 46, 70, 87–8, 104, 122, 159, 242; biomass 17; cells 159; in coastal regions 15; colonies 3, 143, 194–5, 198; cultivation 19; culture 218; cultures 15, 50–1; eukaryotic 17; green 16, 70; growth 50; individual coral polyps hosting (called zooxanthellae) 137; organisms 32; photo-bioreactor 18; photosynthesis 32; photosynthetic 84, 89; species 16–17
microbiome, urban 4, 46, 110, 143, 203
microclimatic effects 143, 262
microorganisms 13, 64, 103, 133, 204, 240
microprocessors 170, 173–4, 198, 242

migrant workers 334
Milan 118, 180–3, 194–5, 199
Milano EXPO2015 196, 198
minerals 16, 22, 24, 263
model 17, 25–6, 115–17, 120, 154, 236, 239–40, 244–6, 311, 315, 319; 3D-printed study 280; abstract 181; advanced 155; algorithmic 186; analogue 249; architectural 117; behavioural 113; bio-mimetic 117; biological 136, 241, 295; computational 245; cyber-Garden 120; digital elevation 324, 327; generative 249, 306; morphological 243; new 79, 180, 253, 255, 267; operational 248; polyp-oriented 115–16; speculative 243; static 237; system 117; systemic 188; virtual 209
modules 76, 84, 86, 143, 146; bio-factory walls 90, 93; integrated photosynthetic walls 90–1; optimising the carbon sequestration process 79
molecules 79, 84, 112; air 28; biotic 263; carbohydrate 17; nitrogen 19; polluting 102
Molinari, Luca 194
monitoring 15, 29, 102–3, 235; controllers 29; devices 102; digital 15; geo-located local 34; high-resolution satellite earth 273; software 19; static local 34; systems 29, 255, 262–3
Montenegro Pavilion *268*, 280
Mori Art Museum, Tokyo 136, *141*, 143
morphogenesis (of micro-algae cells) 54, 115, 137, 248, 277
morphogenetic systems 236
morphologic 50, 55, 116–17, 121, 186–8, 212–13, 236, 242–3, 246, 288, 318–19; effects 288; information 246; traits 116, 121; variations 55
morphology 3, 114–16, 120, 129, 155, 208, 245–8, 275, 278; architectural 99, 222; coral 116; external 154; landscape 183; original 246; of the PhotoSynthetica Tower 143; urban 6, 194, 286, 330
Morton, Timothy 232, 234
mould 113, 240, 244, 246–7; behaviour 247; monocellular 338; searches

246; slime 240–1, 247; tests 247; ubiquitous 3
Museum für Kommunikation, Ausstellung, Frankfurt 160
museums 306; Mori Art Museum, Tokyo 136, *141*, 143; Museum für Kommunikation, Ausstellung, Frankfurt 160; Museum of Future Energy 64, *66*, 69; new Futurium, Berlin 128; ZKM Media Art Museum, Karlsruhe 54

N

natural light 14, 21, 64, 92; *see also* light
natural ventilation 99, 104, 222
navigational mechanisms 298
Navigli (Milan canals) 183
Neom Project, Riyadh 136
Nestle Portuguese, Lisbon 90, *91*
Net Zero Carbon Buildings Commitment 13
networks 3, 5, 154–5, 184, 186, 188, 190, 205, 208, 244–5, 247, 270, 272, 306; biological 205, 296; biotic 333; co-evolving 285, 306; dense 243; exhibition *308*; green 190; green capillary 190; informational 205, 244, 283; minimal 212, 324; municipal waste collection 325, 329–31; mycelium 3; neural 240–1, 340; private transportation 209; re-metabolising Milan's 194; self-aware infrastructural 248; urban wastewater 284, 289
neural networks 240–1, 340
'New Babylon' (Constant Nieuwenhuy) 311, 315
new generations 122, 155, 218
New York Times 236
nitrogen 21, 99, 222; dioxide 32, 35, 99, 222, 314; fixed 22; molecules 19
NO2 see nitrogen dioxide
Nobility House, Helsinki 84–6
Nobility House Urban Curtain 84
nodes 112, 184, 209, 295; blue-green 212; connecting biodiversity 295; multi-species 295
non-humans 155, 300, 302
Normalised Difference Water Index 268, 288
nuclei 245, 247, 314

nutrient regulation 245
nutrients 15–16, 20, 22, 42–3, 68, 116, 120, 245–7, 249, 310–11, 314–15
nutrition 16–17, 44

O

O.serie *157*, *158*, 159, *160*
occupancy sensors 29
oil 23, 28, 70, 181
Open Aviary 270, 272–3, 276–7, *278*, 280
Open Street Maps 31, 35, 327
organisms 15, 17, 114, 181, 188, 233, 239–41, 245, 255, 293, 295; aquatic 20; bio-digital 221–2; biological 122, 187, 242; healthy 118; host 114; human 59; micro-algal 68; mono-cellular 314; multicellular 18; non-human 46; nutritious 54; photoautotrophic 22; semi-autonomous 306; single cell 18, 244; synthetic 122, 143, 236; unicellular eukaryotic 15
Orléans Biennale 2017 254
ornithological parks 265, 267, 273, 280
ornithologists 280
Otto, Frei 188, 232, 243–4
oxygen 17, 27, 34, 47, 65, 68, 132, 142, 193, 226; fresh 42, 102, 220; levels 20; metabolized 218; photosynthesised 28, 79, 146; reduction 34; reduction potential 34; releasing 59, 84, 120

P

Paljassaare Peninsula 282–3, 287–91
paradigm 116, 119, 184, 244; classical 119; embedded technological 184; morphogenetic 117; multi-headed 232; new 118; shift 188
Parco Sud 181, 183
Paris 120, 136, 138, 140, 241, 306, 308–11, 313, 315–20, 339
parks 181, 185, 190, 205
particles 19, 29, 32; liquid medium washes 102; polluting 32; rock 263; small 32
Pask, Gordon 118–19, 155, 239
'Paskian Environments' 119
Pasquero, Claudia 4–5, 19, 46, 62, 103,

117, 123, 136, 186–7, 267, 338–9
Pasteur, L. 236
path systems 188, 190, 244, 306, 327
patterns 15, 29–30, 34, 102, 162, 185, 188–9, 213, 234, 236, 244; average flow 189; building cladding 25; coherent 114; crystallisation 256; curtain 79; digital 242; emerging 189, 262; morphological 50, 186; post-natural 238; serpentine 28; turbulent 258; visible 243; water flow 284, 324
performance 15, 19–20, 65, 114, 185, 195; efficient ecological 218; enhancement 14; environmental 25; evolving system 34; filtering 32; new paradigm 123; problem-solving 238; quantifiable 338; urban 212
PETG see Polyethylene terephthalate petri dish *314*
phaeodactylum tricornutum 16, 70
phagocytosis process 115
pharmaceuticals 16–17, 26
photo-bioreactors 15–16, 18, 25, 27–9, 31, 33, 42, 44, 76, 79, 146, 222, 225; 3D-printed 26, 158; bioplastic 79; glass tube 27; lab grade glass 68; large-scale 26; membrane 25, 27; permanent 25; in photosynthetica 27; single customized 42; technology 25
photosynthesis 17–19, 25, 32, 47, 98, 115–16, 120, 142, 150, 195, 222
photosynthetic 13, 15, 18–19, 78, 132, 214, 270; algal bio-curtains 58; architecture 58; bacteria 21; boiserie 157; cells 22; colonies 65; cyanobacteria cultures 120, 142; efficiency 13, 16; iso-lines 134; microalgae 84, 89; microorganisms 65; organisms 15, 18, 120, 142
PhotoSynthetica 4–5, 12–20, 23–35, 68, 76–7, 79, 99, 158, 222; active ingredient of 30; adoption of 23–4; core process powering 17; cultures in 32; Curtain 75, *77, 78, 80, 81, 85, 86*; Dublin 76; façade 79; Helsinki 38, 84; installing 18; technology 88; TM project 79
PhotoSynthetica Tower 122, 136, 143, *144, 145*, 147
PhotoSyntheticaTM curtains 50
PhotoSyntheticaTM technology 98
Physarum Machines: Computers from Slime Mould 240, 242–4, 246, 248–9
Physarum Polycephalum 233, 240, 242, 244, 249–50, 306, 310–12, 314–15, 318–19
pilot projects 113, 270
PLA see bioplastic
planet 3–4, 112, 118–19, 181, 236–7, 244, 249, 294, 300, 302, 315
PM2.5 pollution monitoring diagrams 108
Poletto, Marco *4*, 5, 19n6, 46, 54, 136, 202, 267, 338–9
pollutants 12, 14–15, 19, 32–3, 99, 222, 236, 277; air 19; and CO2 12; concentrations of 19, 33, 35; metabolising 59, 195; re-metabolizing 16; toxic 13; transformation of waste and 14; urban 32, 35, 220; and waste 14; *see also* air pollution
polycephalum 235–6, 240, 244; aesthetics 233; architecture 232–3, 241
polyethylene terephthalate 142
polyps 115–16, 122, 137
Pont Georges V *256, 257,* 263
portable sensors 35
Price, Cedric 118
printing 26, 142–3, 277, 340; and bio-printing 154; experiments 154; machines 120, 142, 246; technology 26, 54
Printworks Building, Dublin Castle 76–8, 80
problems 21, 123–4, 183, 241, 248, 250, 280; design 239, 249; environmental 182; fundamental 182; human defined 241; human-oriented 250; re-problematising 124
process 17, 19–20, 26–7, 113, 118, 120–1, 123, 154–5, 185–6, 188, 195, 236–7, 241–3, 248, 326–7; abiotic 258; biochemical 276–7; computational 236, 247–8, 340; creative 122; cyber-gardening 120; digestive 115; embryogenetic 234; evolutionary 112–13, 122; flocculation 29; morphogenetic 122, 234; natural 277; organic 302;

production 68, 92; reproductive 114; self-regulating 188
process low-carbon construction 89
processing 65, 69, 92, 212, 240, 277, 282
production processes 68, 92
projects 2–3, 38, 58–9, 98, 120–1, 123, 136, 150–1, 166–7, 194–5, 202, 204, 213, 218, 241–2, 254, 263, 276–7, 298, 302; AirBubble 2, 34, 38, 97–226; Arbor *149*, 150, 156; 'Bio.Tech Hut' 6, 28, 38, 64–6, 68, 70, 73, 120, 146; BioBombola 38–40, 42, *43*, 44; cultural legacy of 202; current 280, 282; design 188, 232, 276; dissertation 340; of ecoLogicStudio 2; Fun Palace 118; GAN-Physarum, and machine vision 34; generator 318; and installations 117; local 334; pilot 113, 270; single 276; sites 35, 257, 277, 283; 'Storytelling Bio.Curtain' 58–60, 62
prokaryotes 114
prokaryotic cyanobacteria 17, 120
proteins 40, 68, 129, 199; and carbohydrates 115; and chlorella 16, 70; meat based 70; vegetable 14, 40, 199
proto-gardens 6, 195, 198
protocols 14, 17, 33–4, 213, 242, 250, 288–9, 318; algorithmic design 117; architectural 122, 124; autonomous farming 89; gardening 30, 118; growth 20, 23, 30; indexical 263; integrated 13; machining production 121; maintenance 30, 34; management 255, 262; satellite monitoring 242
prototypes 63, 113, 166, 187, 204, 278
proximity sensors 29, 164, 173–4, 198

Q

QR codes 30
Queen Elizabeth II 339

R

raw materials 68, 122, 146, 245–6, 334; microalgae based 87; sustainable 88; valuable 12–14

recirculation (air) 22, 99, 104, 222
research 13, 120, 123, 154, 199, 232–3, 239–40, 266–7, 272, 276, 298; academic 2; anthropological 182; Arbor design 154; Bio-digital 122, 146; bio-digital design 47; groups 267; methodology 232; postgraduate 298; practice-based 267; projects 88, 213; strategies 277; teams 267, 282
researchers 3, 30, 121, 240, 267, 282, 340
resolutions 132, 189, 262–3, 268–70, 277, 288, 318, 327; of the 3D printed reactor's cells 159; allowing the formation of triangular bio-pixels 26; of architectural spaces 243; high 26, 190, 311; urban 319
resources 3, 5, 116–17, 121, 151, 247–8, 314
risk management 255
River Loire 262–3
Royal Institute of British Architects Gold Medal award 339; *see also* architects

S

satellites 262, 268, *307, 313, 316, 317, 320, 324, 335, 336*; monitoring the synthetic urban landscape 288; new Sentinella-2 277
sedimentation 256, 258–9
sensors 15, 29–30, 33, 35, 59, 211, 222, 277, 337; aquatic 29; camera 29; environmental 29; light 29; multiple 269; occupancy 29; portable 35; proximity 29, 164, 173–4, 198; urban air pollution 99
Shaviro, Steven 240–2
slime mould 3, 233, 240–2, 244–50, 296, 311, 314–15, 338
software 15, 20, 30
Solana Open Aviary 252, *266*, 268, *271*, 275, 277, *279*, 280
Solana Ulcinj, Montenegro 266–7, 270, 274, 280
solar energy 12, 14, 18, 102–3
solutions 3–4, 12–13, 18, 26, 76, 116, 122, 198–9, 241, 247, 249; advanced 26; bioreactor membrane 18; cost-effective 14; crafting bio-material 12;

design 4; developing architectural 339; economical 25; green 18; innovative 122, 198; spatial 199
spatial intelligence 121
spatial substratum 122, 124
Spirulina 16, 18, 21–2, 24, 28, 42, 90, 198; cells 21, 54; cultivation of edible microalgae 198; cultures 22, 34, 197; in extreme light conditions 21; fresh wet 199; growth 22; growth rate 22, 34; harvesting 40; living photosynthetic strains of 42; nutritional value of 24; platensis 47; risks of damaging 22
Steps to an Ecology of Mind 233
'Storytelling Bio.Curtain' Project 58–60, 62
substratum 137, 142, 188, 246–8, 262–3; augmented 245; biotic 209, 283; digital 244; fabricated 264; growth 247; layered 272; material 248; morphological 242, 246; printed 137; spatial 122, 124; unstable 246; volcanic clay 301
SuperTree 127–9, *130*, 132–3, *134*
Synthetic Landscape Lab, Innsbruck University 23, 46, 58, *60*, 136, 151, 314, 340
synthetic landscapes 4, 23, 46, 58, 60, 136, 143, 151, 288, 314, 340
Systemic Architecture: Operating Manual for the Self-Organizing City 2, 117, 120, 186–7, 339
systems 12, 14–15, 18, 20, 23, 25, 27–31, 33–5, 89, 119–20, 170, 188, 190, 195, 234–7; actuating 30; aeration 27–8, 31–2, 34, 107, 225; air piping 42; algae 28; architectural 26; artificial 270, 306; bio-artificial 149–50; co-evolutionary 5, 234, 237; digital 112, 122, 150; environmental 326; horticultural 68; human 188, 244; inorganic 155, 238; non-human 54, 113, 119, 237; turbulent 180, 185

T

Tallinn 188, 190, 280, 282–3, *284*, *287*, 289
technology 12, 19, 23–4, 28, 58–9, 181, 185, 239, 242, 276, 338; 3D-printing bioplastic 24; analogue 188; current 3D-printing 26; deploying VR 14, 322; designing 12; digital bird-tracking 273; fabrication 15, 25, 154; human 112; impulse welding 25; inflatable 222; innovative software 30; intelligent 155, 337; new 4, 239; photo-bioreactor 25; post-industrial 338; webcam 29
temperature 20–2, 29, 31, 34, 195; cold 21; constant room 314; daily fluctuations 31, 132; freezing 16; maximum welding 25; optimum 22
tendrils 173–6
territories 5, 181, 183–4, 187, 195, 244, 246, 268–9, 272; artificial 273; hostile 300; large 188, 243, 267; peri-urban 181; real world 242; unknown 300; urban 6
thermoplastic polyurethane 25, 219, 222
timber structures 99, 150–1, 153
TPU *see* thermoplastic polyurethane
trees 13, 77, 84, 127–9, 132, 151, 204–5, 208–9, 212, 215; algae-powered artificial 338; large 13, 18, 70, 76, 196, 199, 204; single 208, 212; urban 212; young 42

U

UCL *see* University College London
Ultimaker 3 26–7
UMlab research team 298
University College, London 232, 243, 267, 298, 339–40
University of Innsbruck 2, 5, 23, 60, 136, 156, 266, 314, 339–40
urban agriculture 195, 326, 333–4
Urban Algae Canopy *193*, *194*, *196*, *197*, 198–9
Urban Algae Folly *202*, *203*, 204, *206*, *210*, *211*, 213
urban design 113, 117, 155, 186, 188, 232, 239, 249, 294, 300, 339–40
urban environment 13–14, 19, 30–1, 34, 132, 180, 185, 187, 198, 212, 302
urban infrastructure 112, 122, 124, 180, 189, 205, 212, 243, 248, 255, 263
urban metabolism 195, 281–2
urban microbiome 4, 46, 110, 143, 203
urban systems 6, 245

urban terrain 187, 189
urbanisation 180, 185, 283, 289
urbansphere 112–14, 121, 127–9, 180–1, 184–6, 188, 235–6, 243–5, 248, 280, 305–6

V

vegetable proteins 14, 40, 199
Venice Architecture Biennale 2021 27, 48–9, 53, 56, 266
vertical garden 13, *49*, *50*, 51
Villarreal, Oscar 294, 340
Vucinic, Diana 266

W

Warsaw *61*, 98, *100*, *104*, *105*, 108, 218
waste 4, 14–15, 26, 167, 327
waste infrastructure 329–30, 334–5
Weinberger, Katharina 266–7
Wengrow, David 181–3
wetness 194, 268, 288
wetware 5, 15–17, 20, 30, 187, 190
WHO *see* World Health Organization
WHO AQI index system 33
'wild river' 253–5, 258
wood 150–1, 154, 182; life cycle of 149–50; paste 154; powder 154; species 151; structure 154
World Health Organization 14, 98

X

XL Astaxanthin 120, 137, 143

Z

Zaroukas, Emmanouil 340
Zeitoun, Olivier 306
ZKM Centre for Media Art, Karlsruhe 128

T - #0037 - 180823 - C364 - 254/178/21 [23] - CB - 9780367768041 - Matt Lamination